ESSENTIAL
SCIENCE

THIS IS A WELBECK BOOK

First published in 2020 by Welbeck,
an imprint of Welbeck Non-Fiction Limited,
part of the Welbeck Publishing Group
20 Mortimer Street
London W1T 3JW

A CIP catalogue for this book is available from the British
Library.

ISBN 978-1-78739-446-9

Printed in Dubai

10 9 8 7 6 5 4 3 2 1

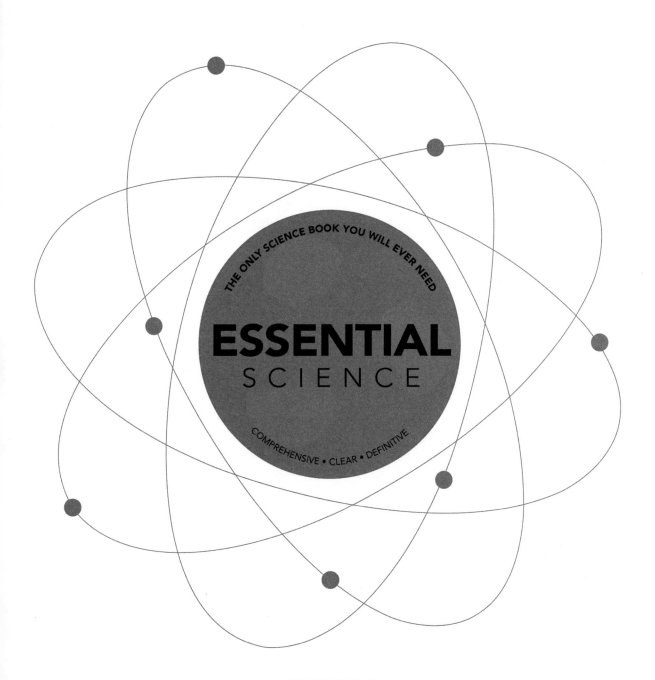

THE ONLY SCIENCE BOOK YOU WILL EVER NEED

ESSENTIAL
SCIENCE

COMPREHENSIVE • CLEAR • DEFINITIVE

BRIAN CLEGG

WELBECK

PHYSICS AND COSMOLOGY

01

CHEMISTRY

02

BIOLOGY AND EVOLUTION

03

EARTH

04

INTRODUCTION

SCIENCE IS CENTRAL TO MODERN LIFE. WE DEPEND ON TECHNOLOGY, MEDICINE AND MORE WITH A BASIS IN SCIENCE TO SUPPORT PRACTICALLY EVERYTHING THAT WE DO. BUT SCIENCE IS FAR MORE THAN THIS: IT IS A RESPONSE TO A FUNDAMENTAL URGE TO KNOW MORE ABOUT THE UNIVERSE WE LIVE IN AND TO UNDERSTAND HOW IT WORKS.

Science is a relatively modern concept, dating back around 2,500 years. In this time, our understanding of everything from the tiniest atom to the universe as a whole has grown immensely. However, this doesn't mean there isn't plenty more still to find out.

The poet John Keats, in his 1819 poem *Lamia* accused science of "unweaving" the rainbow – effectively spoiling nature's beauty by understanding how a rainbow was formed. In reality, though, the reverse is true. We can still delight in the splendours of the universe, but can also get far more out of our experiences by appreciating the remarkable workings of reality.

Inevitably, a book providing an overview of all of science will have omissions. Rather than try to cover everything in sparse detail, I have divided science into four broad sections, and within each picked out a number of essential topics. The sections are physics and cosmology, chemistry, biology and evolution, and earth science.

By keeping to a relatively small number of topics – 34 in all – we can explore in more detail, starting each topic with a concise summary, moving on to see where the concept came from in history, looking at the key theories and evidence, how the idea has been criticized in the past, why this specific topic matters and how it might develop in the future. Each topic finishes with a summary table to pull it together. This structure both makes the science accessible and emphasizes the way that science is a never-ending process, constantly reshaping our perceptions of reality.

We begin with the central mystery that makes everything work. What is energy?

Opposite: Orion Nebula is the middle star in the handle of the pot in the constellation of Orion, 1500 light years away.

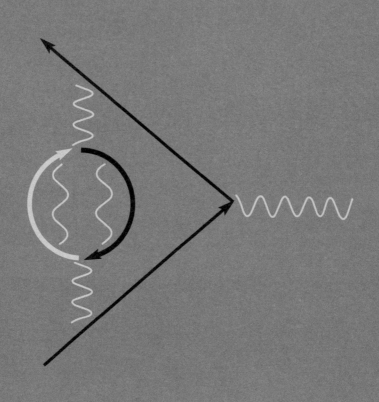

PHYSICS AND COSMOLOGY

WORK, POWER AND ENERGY

THE **ESSENTIAL** IDEA

"NOTHING CAN BE LOST IN THE OPERATIONS OF NATURE –
NO ENERGY CAN BE DESTROYED."

WILLIAM THOMSON (LORD KELVIN), 1849

Three strongly related concepts are at the heart of making things happen in the physical world: work, power and energy. In normal English use, these are terms that are almost interchangeable, but they have very specific meanings for scientists. Although it is extremely difficult to describe what we mean by energy, it has the responsibility for making change happen.

Energy comes in a range of forms – for example, the potential energy provided by gravity or a spring, chemical energy, electromagnetic energy or the kinetic energy of movement and heat. Within a closed system, where energy cannot enter or leave, the amount of energy always stays the same, though it can be converted from one form to another.

Work is, effectively, what energy does. It refers to energy being transferred from one place to another or changing form. So, for example, we "do work" when we lift a heavy object off the ground, because we are transforming chemical energy from our body into the potential energy that the heavy object now has being higher up in the Earth's gravitational field, plus heat energy generated by our body.

Power is simply the rate at which work is done. In science, energy (and hence work) is measured in joules and power in watts (joules per second).

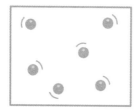 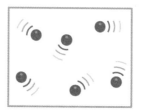

Top: Electrical energy is converted to kinetic energy and heat by the train's motors.
Bottom: Temperature reflects the kinetic energy of atoms or molecules: the faster the movement, the greater the energy.

ORIGINS

The scientific concept of energy is a relatively modern one. Although natural philosophers from the ancient Greeks to Isaac Newton studied motion and the impetus for an object to stay in motion – what we now call momentum – energy itself came significantly later. Of course, aspects of energy such as heat were recognized, but there was no linking theme to pull the concepts together.

The first step in the establishment of energy as a scientific term came from Newton's contemporary, the German mathematician Gottfried Leibniz, who in the seventeenth century described "vis viva" (meaning "living force" in Latin) that was approximately what we would now call kinetic energy. The word "energy" only came into use for this concept in 1802, when it was used by English scientist Thomas Young in a lecture to the Royal Society in London. The word itself dates back to the sixteenth century, but only in the general way it is used in everyday English.

"THE WORD 'ENERGY' ONLY CAME INTO USE FOR THIS CONCEPT IN 1802, WHEN IT WAS USED BY ENGLISH SCIENTIST THOMAS YOUNG IN A LECTURE TO THE ROYAL SOCIETY IN LONDON."

That first mention of energy only referred to motion – kinetic energy. It wasn't until 1853 that Scottish engineer William Rankine described potential energy, with other forms being added in later as a better understanding was developed of physical nature. A particularly important development was the understanding of heat as a form of energy.

THEORIES OF HEAT

In the late eighteenth century, the theory of caloric was developed. This described heat as an invisible fluid which flowed from things that were hot to things that were colder. This meant that heat was conserved – it could flow from place to place but could not be made or destroyed. Though incorrect, caloric theory was surprisingly successful and would be used by nineteenth-century French scientist Sadi Carnot to develop an explanation of how steam engines worked, providing the basis of the new science of thermodynamics.

However, heat's time as a strange fluid was running out. In the 1840s, the English brewer and scientist James Joule was investigating the economics of different ways to power machines in the brewery and discovered a relationship between heat and mechanical work. He used a number of devices to demonstrate this, the best known being one that used a falling weight to power a paddle spinning in water, where the potential energy lost was equated to the increase in heat in the water.

Top: Sadi Carnot (1796–1832).
Bottom: James Joule (1818–1889).

KEY **THEORIES** AND **EVIDENCE**

NATURE, CONSERVATION AND SYMMETRY

Energy comes in a wide range of guises, though some are more basic forms in disguise. The simplest type to understand is kinetic energy – the energy of movement. Dependent solely on the mass of a body and the square of its velocity, kinetic energy was the first type of energy to be identified as such.

Another familiar form is potential energy. Potential energy is stored-up energy, not causing any change, but capable of doing so. Most easily recognized is gravitational potential energy. Gravity is an interaction between two objects with mass that causes them to be attracted to each other. This means that it takes energy to move one object away from another, stored as potential energy. Gravity pulls the objects towards each other: when the object is allowed to move, the gravitational potential energy is converted first into kinetic energy and then into heat and sound energy when the objects collide.

TYPES OF ENERGY

Although gravitational potential energy is the most obvious, there are plenty of other ways for energy to be stored. For example, a coiled spring has potential energy, as does a stick of dynamite or a piece of radioactive material. Such examples of potential energy are often electromagnetic at their heart. To see why, we first need to identify what electromagnetic energy is.

Electromagnetic energy refers to the interaction between electrical charges and between magnetic poles. Electrical charges and magnetized objects attract or repel each other, producing similar effects to gravity. Gravity is relatively weak – the gravitational attraction between two atoms is hardly noticeable – but the electrical or magnetic attraction (or repulsion) has a significant effect. In the examples given above, bonds in a coiled spring – the attraction between atoms – are stretched, storing energy in a similar way to lifting an object away from the Earth. An explosive such as dynamite gets its energy from the potential energy in chemical bonds which are broken. And radioactive energy comes primarily from the potential energy tying the nuclei of atoms together.

"GRAVITY IS AN INTERACTION BETWEEN TWO OBJECTS WITH MASS THAT CAUSES THEM TO BE ATTRACTED TO EACH OTHER. THIS MEANS THAT IT TAKES ENERGY TO MOVE ONE OBJECT AWAY FROM ANOTHER, STORED AS POTENTIAL ENERGY."

Another familiar form of electromagnetic energy is light. The interaction between electricity and magnetism can provide a transmittable kind of energy, which can be considered as a light wave or a stream of photons. Most of the energy we use on the Earth reaches us in light from the Sun.

We have already mentioned heat. Although it is often convenient to treat heat as a thing in its own right, in reality it is primarily kinetic energy – the energy of motion of the atoms in a physical substance, whether it is a solid, liquid or gas.

Whatever its form, energy is conserved. It can be moved from place to place and converted between forms, but it cannot be created or destroyed. Clearly a body that is connected in some way to external objects can gain or lose energy. We gain energy from our food and lose it as heat and in performing physical tasks. However, a body that is totally isolated, known as a closed system, cannot gain or lose energy and it is in this circumstance that energy is conserved.

EVIDENCE

We see energy at play in a wide range of the sciences, from the energy of stars in cosmology through to the biological interplay of energy involved in life. Once energy was established as a concept in the nineteenth century, it was through experiments such as Joule's, demonstrating the transfer of potential energy into heat, that energy values began to be quantified. Understanding heat energy was particularly important in the nineteenth century with its dependence on steam engines, but other aspects of energy conversion came to the fore in the twentieth century.

We tend to refer loosely to energy production or energy sources (for example, in a power station), but energy is never produced: instead it is converted from one form to another and transferred from place to place. Most of the energy we use on Earth comes from the Sun, converted by plants into chemical energy that enters the food chain as well as warming the planet. A portion also comes from the Earth itself, where inner radioactivity generates heat. All these energy-transfer mechanisms have been studied and provide evidence that supports our understanding of energy.

Top: In the power station energy is converted from one form to another.
Opposite: Explosives release potential energy from chemical bonds.

13

CRITICS

The aspect of energy that has caused most controversy over the years is its conservation. For hundreds of years it was thought that it should be possible to build a "perpetual motion" machine. This is one that can keep moving forever once it has been started without any further input of energy. Although in principle this seems possible, in practice all machines have some inefficiencies – they lose some of their energy to heat through friction and air resistance.

Perpetual motion machine designs often involve a device that rotates, where the process of rotation causes something to happen (perhaps liquid or magnets moving from place to place) which sustains the motion.

> "MOST PROMINENT AMONGST RECENT FAILURES OF PERPETUAL MOTION MACHINES WAS A DEVICE BUILT BY IRISH COMPANY STEORN IN 2006."

Most prominent amongst recent failures of perpetual motion machines was a device built by Irish company Steorn in 2006. With much publicity they claimed that they could produce "free, clean and constant energy" from a small device that used rotating magnetic fields to generate energy. Despite an attempt at a public demonstration, they were never able to prove that their device functioned.

Some claim that we should be able to access "zero-point energy", which is a base level of energy in the universe that is required by quantum theory. Unfortunately, to make use of a source of energy we need somewhere else with lower energy so that the energy can do work by flowing from one to the other. As zero-point energy is, by definition, the lowest possible, it cannot be used.

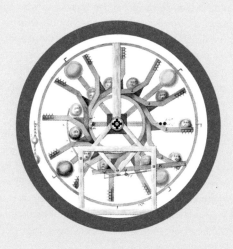

WHY IT **MATTERS**

"IT IS IMPORTANT TO REALIZE THAT IN PHYSICS TODAY,
WE HAVE NO KNOWLEDGE OF WHAT ENERGY *IS*."

RICHARD FEYNMAN, 1964

Energy is fundamental to everyday life. Although it is, as famous American physicist Richard Feynman pointed out, pretty much impossible to actually say what energy *is*, its importance for us is what it *does*. Energy drives life. Energy is what makes things happen. And as we now realize that matter and energy are different aspects of the same thing, energy is not just what makes things happen, it is what things are made of in the first place.

Today there is a lot of concern about the different ways that we generate energy. (Or, more precisely, how we convert energy into more usable forms.) We talk about moving from fossil fuels, for example, to "renewable energy". In practice this is a label that has little scientific merit. You can't "renew" energy. If you consider the system of the Sun and the Earth – the system that is responsible for our lives past, present and future – there is a fixed amount of energy available. This is being pumped out by the Sun and can be temporarily stored in various potential forms, such as the chemical energy in plants which eventually become fossil fuels.

Many human problems come down to needing energy to be available in the right place at the right time. Nothing is more fundamental to humanity than energy.

Above: Richard Feynman (1918–1988).
Opposite: Design for a gravitationally based perpetual motion machine.

FUTURE **DEVELOPMENTS**

It is unlikely that we will discover much that is particularly new about energy itself. The basics are amongst the best understood aspects of science (even if we don't strictly know what energy is). Developments in the field of energy are most likely to come in new and more effective ways to transfer energy from one type to another and to store it away.

As we move energy generation more and more from fossil fuels to sources powered by the Sun and nuclear energy, developments are often in the efficiency of the conversion processes and storage. So, for example, a key aspect of energy transference is from the electromagnetic energy of sunlight to chemical and electrical energy. This happens in nature in the process of photosynthesis, where plants turn sunlight into chemical energy. We make use of photovoltaic solar cells, which convert light energy directly into electrical energy, or more indirect sources, such as wind, where the sunlight heats the air, causing it to move, then the moving air is used to turn turbines which generate electricity. Such devices will never be 100 per cent efficient, but we are making huge advances in their effectiveness.

Above: A solar farm using photovoltaic cells.

THE **ESSENTIAL** SUMMARY

ORIGINS	KEY THEORIES AND EVIDENCE	CRITICS	WHY IT MATTERS	FUTURE DEVELOPMENTS
1695 Gottfried Leibniz describes "vis viva", an early equivalent of kinetic energy.	Energy comes in a number of forms, notably **kinetic energy** and **potential energy**.	Early energy theory was criticized by those who supported the **caloric** idea that heat was a fluid transferred from hotter to colder bodies – but this criticism soon faded.	Although we don't know what it *is*, **energy is fundamental to life** itself.	Development is in making **energy generation** (transfer) **more efficient** and in achieving better **storage** of energy.
1802 Thomas Young makes the first use of the word "energy" in the modern sense.	Other examples of forms of energy include **gravitational, electromagnetic, chemical, nuclear** and **heat**.	Most critics of conservation of energy believe it is possible to create **perpetual motion machines** – but no such device has been successfully created.	Energy is what **makes everything happen**.	A crucial aspect of energy transfer is **light from the Sun** being **converted** into electrical and chemical energy – greater efficiency here is crucial.
1843 James Joule reports on his first experiments showing the relationship between kinetic energy and heat.	Most of the energy used on the Earth comes from the **Sun** as electromagnetic energy.	Some have suggested perpetual motion is possible by harnessing the universe's **zero-point energy** – but this is not possible.	Matter and energy are **equivalent** – so energy is also what everything is made of.	Energy conversion and storage can **never be 100% efficient** – but we have a long way to go.
1853 William Rankine first describes the concept of potential energy.	**Energy is conserved.** It can be transferred from place to place and form to form, but it cannot be created or destroyed.		"Energy generation" is really just about **transfer of energy** from one form to another.	

THERMODYNAMICS

THE **ESSENTIAL** IDEA

"THE SECOND LAW OF THERMODYNAMICS HAS THE SAME DEGREE OF TRUTH
AS THE STATEMENT THAT IF YOU THROW A TUMBLERFUL OF WATER INTO THE
SEA YOU CANNOT GET THE SAME TUMBLERFUL OF WATER OUT AGAIN."

JAMES CLERK MAXWELL, 1870

Although it started as the science required to explain the working of steam engines, thermodynamics expanded from being an exploration of the movement of heat to one of the most fundamental aspects of science, explaining why time has an apparent direction and predicting the way that the universe will end.

At the heart of thermodynamics are four laws which deal with the way that heat flows, the conservation of energy and the lowest limit of temperature, absolute zero. We have already seen the importance of conservation of energy (the first law), but arguably even more important is the second law.

Taken simply, the second law says that, without the input of energy, heat will move from a hotter place to a colder place. An alternative formulation is that in a closed system, entropy will stay the same or decrease. Entropy is a measure of the disorder in a system – the bigger the entropy, the more the disorder. In effect, this means that the second law explains why it is easier to break a glass than to unbreak it.

Above: The second law says that heat moves from
hotter to colder and disorder increases.
Opposite top: The distribution of velocity of molecules
in a gas corresponds to temperature.

ORIGINS

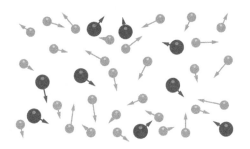

Not surprisingly, it was in the nineteenth century, when the industrial revolution was being driven by the steam engine, that thermodynamics took off as a branch of physics. It was the science that provided a better understanding of how heat produced work. The first major figure involved in developing thermodynamics was French physicist Sadi Carnot, who wrote a book called *Réflexions sur la Puissance Motrice du Feu* (Reflections on the Motive Power of Fire). Despite being based on the soon-to-be-dismissed caloric theory, which regarded heat as a fluid, Carnot was able to identify some key aspects of thermodynamics, notably that the efficiency of heat engines such as steam engines was dependent on the difference in absolute temperature between the hot part and the cold part of the system. He also made it clear how important it was to deal with a closed system – if there were external influences, it would change the workings of the system.

Many of the key figures in the development of thermodynamics during the 1850s came from Scotland, a country that was very active in engineering and applied physics at the time. Key figures included the engineer William Rankine and physicists James Clerk Maxwell and William Thomson, Lord Kelvin. They would be joined by German physicist Rudolf Clausius and Austrian physicist Ludwig Boltzmann in the clarification of much of the basic theory of thermodynamics.

The contributions made by Maxwell and Boltzmann were particularly important in that they made it clear that thermodynamics was a statistical discipline. At the time, there was still significant dispute over whether or not atoms and molecules existed. Maxwell and Boltzmann showed how temperature was dependent on the distribution of velocities of molecules in gases, with faster molecules corresponding to higher temperatures. With vast numbers of atoms present, the effect could not be worked out in detail: it had to be treated statistically.

Apart from providing an explanation for the operation of thermodynamics, the statistical approach demonstrated that, unlike other physical laws, some of the mechanisms of thermodynamics only told us what was most *likely* to happen, rather than describing an absolute certainty.

Top right: William Rankine (1820–1872).
Bottom: James Clerk Maxwell (1831–1879).

KEY **THEORIES** AND **EVIDENCE**

LAWS, TEMPERATURE AND ENTROPY

"IF SOMEONE POINTS OUT TO YOU THAT YOUR PET THEORY OF THE UNIVERSE IS IN DISAGREEMENT WITH MAXWELL'S EQUATIONS – THEN SO MUCH THE WORSE FOR MAXWELL'S EQUATIONS. IF IT IS FOUND TO BE CONTRADICTED BY OBSERVATION – WELL THESE EXPERIMENTALISTS DO BUNGLE THINGS SOMETIMES. BUT IF YOUR THEORY IS FOUND TO BE AGAINST THE SECOND LAW OF THERMODYNAMICS I CAN GIVE YOU NO HOPE; THERE IS NOTHING FOR IT BUT TO COLLAPSE IN THE DEEPEST HUMILIATION."

ARTHUR EDDINGTON, 1927

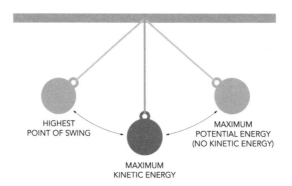

HIGHEST
POINT OF SWING

MAXIMUM
POTENTIAL ENERGY
(NO KINETIC ENERGY)

MAXIMUM
KINETIC ENERGY

At the heart of thermodynamics are four laws. These start, confusingly, with the zeroth law – so named because it is the most fundamental, but was late in being established. It gives us the concept of an equilibrium – a balanced state – as far as heat is concerned, telling us that two objects are in equilibrium if heat can flow between them, but the net flow of heat is zero. This "net flow" part is important because of the statistical nature of thermodynamics. There will be heat flowing in both directions, but overall, averaged out, the zeroth law tells us there will be no flow.

The first law we have already met. In thermodynamic terms it is sometimes phrased as "the energy in a system changes to match the work it does on the outside – or the outside does on it – and the heat that is given out or absorbed." In effect, the amount of energy in a system doesn't change unless it comes in from the outside or is lost. In an isolated system, energy is conserved.

The second law says that in a closed system heat will move from a hotter part of the system to a cooler part – or that entropy stays the same or increases. Entropy is the measure of the disorder in a system, formally quantified as the number of different ways the components of a system can be organized. If you have a set of alphabet blocks, one for each of the letters of the alphabet, and have them all in alphabetical order, this arrangement (alphabetically speaking) has low entropy, because there is only one way to do it. But scramble them up and the entropy increases, as there are many different ways to have the blocks scrambled. It takes less work to go from a low-entropy arrangement to a scrambled arrangement

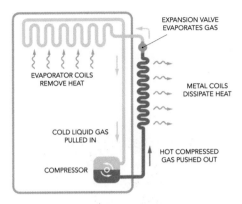

than the other way. The same applies to breaking something, which is why the second law is sometimes typified as "everything wears out".

The second law is not irreversible as it is rare in the real world to have a closed system. For example, the existence of living things is an example of entropy being reduced on the Earth. However, the Earth is not a closed system – a vast amount of energy comes in from the outside in the form of sunlight. Most of us have a good example of reversing the second law: a refrigerator. Here heat moves from a colder place (the inside of the fridge) to a warmer place (the room outside). Again, this happens because energy is pumped into the system.

The third law says that it is not possible to get a body down to absolute zero (0K, –273.15°C or –459.67°F) in a finite number of steps. Absolute zero is the lowest possible temperature. Bearing in mind that temperature is a measure of the kinetic energy of atoms and molecules (along with the energy of electrons within the atoms), the third law tells us there is a point at which a body is at its lowest energy level, but which can never be practically achieved.

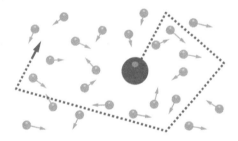

EVIDENCE

Thermodynamic theory depends on the existence of atoms, which were shown to exist theoretically by Einstein in 1905 and soon after experimentally. The third law was demonstrated originally in the early 1900s from statistical measures, but would become definitive when quantum physics was developed. The zeroth law was added in the 1930s as the formalization of a long-established reality.

Opposite: A pendulum demonstrates the first law, transferring energy between potential and kinetic.
Top: A refrigerator can transfer heat from a colder to hotter location as it is not a closed system.
Bottom: Einstein theorized that collisions with molecules caused the random motion of small particles.

CRITICS

Most of the criticism of thermodynamics came from its statistical nature and its implications for the nature of reality. Opposition from leading European scientists was both to the existence of atoms and the requirement that, because the second law is statistical, sometimes a body would spontaneously go against the law. Some have suggested that this fervent opposition contributed to Boltzmann's suicide in 1906.

The best known criticism is of the second law itself, coming from Maxwell, one of the architects of thermodynamics, in the form of an imaginary entity called "Maxwell's demon".

Maxwell imagined a pair of boxes containing gas at the same temperature with a door between them. When we say the gases are the same temperature it refers to the average speed of the molecules – but some will be faster and some slower. Maxwell imagined a tiny being, the demon, which opened and closed the door when molecules approached, only letting fast molecules travel from left to right (say) and slow from right to left. As a result, heat would flow from the cooler to the hotter side and the level of entropy would decrease, as it is more ordered to have all the hotter molecules in one box and all the cooler molecules in the other. But the demon could achieve this without any input of energy.

The demon was debated for decades, and it still demonstrates well the statistical nature of the second law.

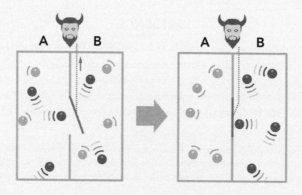

WHY IT **MATTERS**

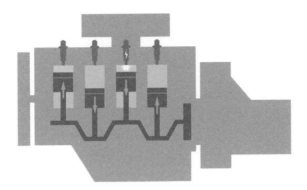

Heat engines are extremely important to our world – no longer in the form of steam engines but as internal combustion engines and ways of generating electricity. More crucial, though, is the second law's influence on reality and our future.

Generally, the laws of physics are reversible – they run equally well in either direction in time. There is no real sense of past and future in most of the laws of physics. But the second law of thermodynamics is an exception. The requirement for entropy to stay the same or increase provides an "arrow of time" pointing to the future. It is argued that much of our sense of progress into the future is driven by the second law.

CLOSED SYSTEMS

As we have seen, we rarely see a closed system in nature. However, we can regard the universe as a whole as a closed system. This means that the second law tells us that, though there may be local statistical reversals, over time, the entropy of the universe will increase. In the long term, it tells us that, unless there is an influence from outside the universe as we know it, eventually everything in the universe will run down, resulting in the so-called "heat death", where nothing notable happens. We are looking many billions of years into the future for this scenario to occur – but that is what the second law predicts for us.

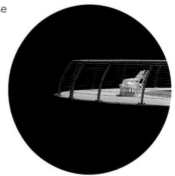

Opposite top: Ludwig Boltzmann (1844–1906).
Opposite bottom: Maxwell's demon opens and closes the door to separate fast and slow molecules.
Top: An internal combustion engine, such as a petrol car engine, is a heat engine.
Bottom: Douglas Adams' fictional Restaurant at the End of the Universe.

FUTURE **DEVELOPMENTS**

"SUPERCONDUCTING MAGNETS ARE USED IN PARTICLE ACCELERATORS SUCH AS THE LARGE HADRON COLLIDER AT CERN"

Although the basics of thermodynamics are established, the biggest developments are at extremely low temperatures. Near to absolute zero, where energy levels are very low, materials begin to act very differently. This is the result of quantum effects, only made possible because of the thermodynamic implications of the extremely low temperatures.

A key area here is superconductivity. Take a conducting material to a low enough temperature and it loses all electrical resistance. This means that there is no heat loss, as there is normally when electricity flows through a wire – and specialist electronic applications become possible. Superconductivity also enables far more powerful electromagnets to be produced than would otherwise be possible. Superconducting magnets are used in particle accelerators such as the Large Hadron Collider at CERN, near Geneva, in MRI scanners and in magnetic levitation trains, which float above the track to enable speeds of over 350 mph (560 kph) to be achieved.

Some of these extremely low-temperature experiments explicitly make use of Maxwell's demon-type effects to manipulate the second law of thermodynamics to make new outcomes possible.

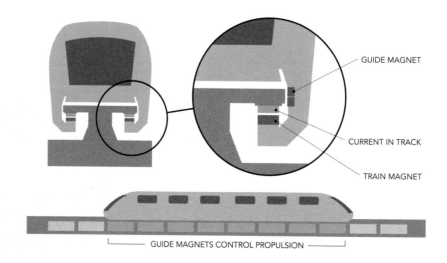

GUIDE MAGNET

CURRENT IN TRACK

TRAIN MAGNET

— GUIDE MAGNETS CONTROL PROPULSION —

Above: Magnetic levitation trains use superconducting magnets both for propulsion and to float over the track.

THE **ESSENTIAL** SUMMARY

ORIGINS	KEY THEORIES AND EVIDENCE	CRITICS	WHY IT MATTERS	FUTURE DEVELOPMENTS
1824 Sadi Carnot publishes his book *Réflexions sur la Puissance Motrice du Feu*, which establishes the basics of thermodynamics. **1850s** First and second laws of thermodynamics established from the work of William Rankine, William Thomson and Rudolf Clausius. **1850s–1860s** Statistical nature of thermodynamics established by James Clerk Maxwell and Ludwig Boltzmann. **1905** Albert Einstein theoretically proves the existence of atoms.	**Zeroth law** says that two bodies are in equilibrium if the heat can flow between them, but the net flow is zero. **First law** says that energy is conserved in a closed system. **Second law** says that heat moves from hotter to colder bodies in a closed system – also phrased as entropy stays the same or increases. **Third law** says that absolute zero cannot be reached in a finite number of steps. Evidence came from **experiments on heat** and the ability to reach **ultra-low temperatures**.	Early opposition was based on doubts about the **existence of atoms** and the **statistical nature** of the second law. An apparent breach of the second law in **Maxwell's demon** would challenge scientists for decades, but fundamentally demonstrates that the statistical nature of the second law is true.	**Heat engines**, understood through thermodynamics, are essential to the modern world in car engines and electricity generation. The second law gives us the **arrow of time**, underlying our understanding of past and future. Assuming the universe is a closed system, the second law tells us the universe will run down eventually to a **heat death** where nothing further happens.	Low-temperature physics, near to **absolute zero**, has major possibilities for new technologies.

CLASSICAL MECHANICS

THE **ESSENTIAL** IDEA

"EVERY BODY PERSEVERES IN ITS STATE OF BEING AT REST OR OF MOVING UNIFORMLY STRAIGHT FORWARD, EXCEPT INSOFAR AS IT IS COMPELLED TO CHANGE ITS STATE BY FORCES IMPRESSED."

ISAAC NEWTON, 1687

- AN OBJECT CONTINUES IN ITS STATE OF REST OR MOTION UNLESS AN EXTERNAL FORCE IS APPLIED TO IT

- THE GREATER THE MASS OF AN OBJECT, THE GREATER THE AMOUNT OF FORCE IS NEEDED TO ACCELERATE IT

- FOR EVERY ACTION, THERE IS AN EQUAL AND OPPOSITE REACTION

FORCE → ← MOTION

It's easy to think of mechanics as the boring bits of physics we spend far too long on in school – how objects move, forces and machines. Yet it was probably the first aspect of physics to be seriously considered in a scientific fashion and is behind everything from the simplest physical activity to what keeps an airplane in the sky.

At the heart of our understanding of mechanics are Newton's laws of motion, and though these would need a modification provided by the special theory of relativity, the seventeenth-century version continues to be all we need for everyday use.

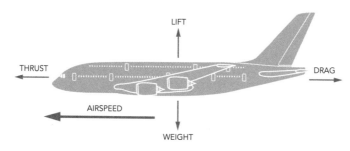

Although mechanics is sometimes limited to bodies in motion, it makes sense to include simple machines, most known since ancient times, and typically used to turn a relatively small force into a larger one. The classical simple machines are the lever, the wheel, the pulley, the inclined plane, the wedge and the screw.

ORIGINS

Ancient Greek mechanics started with the work of Aristotle in the fourth century BC. He envisaged two different kinds of motion – natural and unnatural. Natural motion reflected the tendencies of motion of the elements that Aristotle's physics was based on – earth, air, fire and water – to move towards or away from the centre of the universe. Earth had the strongest tendency towards the centre, water somewhat less so, air a tendency to move away from the centre and fire the strongest tendency to head outwards. Unnatural motion was that imposed on bodies – for example when we push something and it moves – but Aristotle assumed a body needed a constant push to keep it in motion.

Although Aristotle's theories were largely accepted through to the Renaissance, there was some earlier questioning of his mechanical ideas, both by Islamic scholars and English philosophers. But it would take Galileo Galilei to make the first major inroads into turning around Aristotle's ideas. In his masterpiece *Discorsi e dimostrazioni matematiche intorno a due nuove scienze* (Discourses and Mathematical Demonstrations Relating to Two New Sciences), published in 1638 after his house arrest, Galileo studied the movement of pendulums and of objects rolling down an inclined plane. He noted, for example, that an object rolling downhill would accelerate and one rolling uphill would decelerate – so it seemed logical that on the flat it would do neither, whereas Aristotle assumed it would naturally stop.

Galileo's ideas would be crowned by Isaac Newton in his central work on mechanics and gravity, the *Philosophiæ Naturalis Principia Mathematica* (Mathematical Principles of Natural Philosophy), published in 1687. Generally known as the *Principia*, this book contains both Newton's laws of motion and his mathematics of gravitational attraction, establishing the ultimate picture of mechanics until the twentieth century.

Opposite top: Newton's laws of motion.
Opposite bottom: The forces acting on a aircraft in flight, based on Newton's laws.
Top: Aristotle (384–322 BCE).
Bottom: Isaac Newton (1643–1727).

KEY **THEORIES** AND **EVIDENCE**

NEWTON'S LAWS AND SIMPLE MACHINES

Newton's laws of motion are simple yet powerfully descriptive of how objects move. In assembling them, Newton saw past the limitations of the everyday world, where friction and air resistance meant that objects appeared to slow down of their own accord, to realize that this slowing is the result of a force acting on them. His three laws were:

1. A body stays at rest or moves uniformly in a straight line unless a force acts on it.

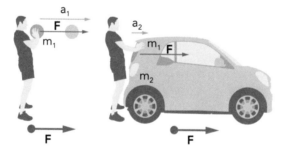

2. A change in motion is proportional to the force applied and takes place along the straight line along which the force acts.

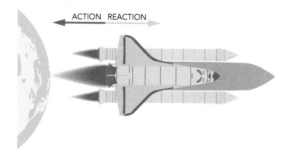

3. Every action has an equal and opposite reaction.

Unlike Aristotle's view that an object needs to be constantly pushed to keep it moving, Newton realized that an object will keep moving at the same speed (or remain still) until something acts on it. That "something" could be a whole range of things: a push, friction slowing it down, or gravity acting on it, for example. Once something is moving it has inertia: it keeps going until something stops it.

Although Newton was the first to use the term "mass" to describe a body's inherent matter content (as opposed to weight, which varies dependent on the strength of gravity), he did not use it in his original statement of the second law. We are now more likely to say that the force applied is equal to the mass of the object times the acceleration that is produced by that force: in equation terms, $F = ma$.

THE THIRD LAW

The third law is the least obvious in its wording, because it seems to imply that nothing would ever move, because there would be an action and an equal, opposite reaction. Newton felt the need to add "in other words, the actions of two bodies upon each other are always equal and always opposite in direction", emphasizing that the action and the reaction are on two different bodies.

We feel the third law when we push a heavy object and it resists that push. It's particularly obvious when a plane takes off. The engines work by pushing air out of the back of the jet – the opposite reaction is the engine being pushed forward, taking the plane with it. As the plane accelerates, the seat is pushed into your back: you feel the reaction of your back pushing into the seat. And the way that the wings keep the plane in the air is in part due to the third law: the wing is shaped to push air downwards, which pushes the wing up.

When we use simple machines, they reflect a simple relationship between force and work. The work required to move something (ignoring friction) is the force applied times the distance over which the object is moved. What the machine does is to apply the work over an extended distance, magnifying the force. (It could reduce the force, but this is rarely desirable.) So, for example, when a pulley system is used, the person hauling on the rope pulls the rope a greater distance than the object being lifted is moved. As a result, they can lift a greater weight. The same applies to other simple machines such as a lever.

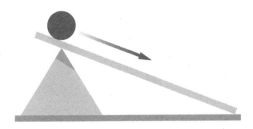

EVIDENCE

Early evidence came from simple experiments by Galileo, often experimenting with the force of gravity in a controlled way. Rather than drop balls off the leaning tower of Pisa as legend suggests he did, Galileo observed pendulums and rolled balls down slopes. Newton's laws would be verified by reducing friction to a minimum (for example, using ice or air tables) and by measuring forces using calibrated devices such as spring-based scales.

CRITICS

Inevitably there was some resistance to Galileo's ideas from those who were set in Aristotelian ways, just as there was resistance to moving away from Aristotle's Earth-centred universe to the Copernican Sun-centred viewpoint. In fact, the two issues were more strongly tied together than is obvious. Although there were no religious objections to Galileo's study of motion, Aristotle's theory of motion was strongly tied to his Earth-centred view of the universe, because he believed that gravity was a tendency of certain elements to move towards the centre of the universe. Even so, Galileo's careful experiments on motion were hard to counter.

"EVERY NOW AND THEN WE DO GET AN APPARENT CRITICISM OF NEWTON'S LAWS IN THE SUGGESTION THAT SCIENCE CANNOT EXPLAIN HOW A PARTICULAR EVENT IS POSSIBLE"

Newton also received a degree of resistance to his ideas, but this was more down to the concept behind his work on gravity than the laws of motion. Every now and then we do get an apparent criticism of Newton's laws in the suggestion that science cannot explain how a particular event is possible, perhaps most famously the claim that the bumblebee should not be able to fly. In reality this is a fallacy (based on a sermon), rather than a scientific argument. Bumblebees' wings don't work like bird wings, but more like helicopter blades, producing a much stronger lifting effect than would be the case by simply flapping.

WHY IT **MATTERS**

*"NATURE AND NATURE'S LAWS LAY HID IN NIGHT.
GOD SAID, LET NEWTON BE! AND ALL WAS LIGHT."*

ALEXANDER POPE, c1727

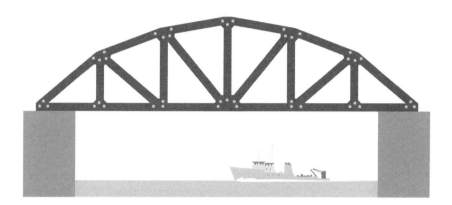

From the earliest use of tools, whether it was hand axes or the first rollers and wheels, we have been making use of basic mechanics. An understanding of this was absolutely essential once our machines and buildings became more sophisticated. While early developments were possible with trial and error, we now would not build a bridge or a building – or devise any kind of mechanical device – without getting a clear understanding of the forces and movements involved in making it work. Engineering as a discipline is entirely based on Newton's laws.

It's not just a matter of the constructed world, of course, as the quote from Pope above, suggested as an epitaph for Newton's tomb, makes clear. Newton's laws of motion are just as relevant to understanding how a bird can fly or how a human being can lift an object off the ground. Almost everything in the universe moves and interacts with other objects around it and Newton's laws give us a clear understanding of how these movements and interactions take place.

Opposite: The bumblebee's wings work differently to a
bird's, but still obey Newton's laws.
Above: Understanding the forces at play in a bridge are
essential for its safe construction.

FUTURE **DEVELOPMENTS**

"KANGAROOS WERE ONCE THOUGHT TO BREAK THE SECOND LAW AS THEY SEEM TO PRODUCE MORE ACCELERATION THAN THEIR INTAKE OF ENERGY WOULD ALLOW."

The basics of mechanics have not changed hugely in several hundred years for a good reason – they are entirely effective. Newton's laws only lose their accuracy when objects are moving at extremely high speeds, when Einstein's special theory of relativity takes over. This is a modification of Newton's laws that takes into account the special nature of light and the implications of what it shows us for the relative nature of time and space – which results in the need to make some changes to the outcome of Newton's laws.

What we tend to find instead is that we continue to fill in gaps of understanding where phenomena apparently challenge Newton's laws. So, for example, kangaroos were once thought to break the second law as they seem to produce more acceleration than their intake of energy would allow. But the disparity was found to be that, like a rubber ball, they absorb some energy when they land which is used to accelerate them away from the ground, using significantly less of the animal's energy store.

Above: Despite appearing to use more energy than they consume, kangaroos don't break Newton's laws.

THE **ESSENTIAL** SUMMARY

ORIGINS	KEY THEORIES AND EVIDENCE	CRITICS	WHY IT MATTERS	FUTURE DEVELOPMENTS
4th century BC Aristotle theorizes that bodies had a tendency to move to their natural location and that it otherwise took a push to keep things moving.	**Newton's First law** says a body stays at rest or in constant motion unless a force acts on it.	Galileo had some resistance from those who still supported **Aristotle's theories**.	Many **tools and machines, buildings and bridges** depend in their design on Newton's laws.	The basics don't change, but when objects are moving extremely quickly, Newton's laws have to be modified with Einstein's **special theory of relativity**.
1638 Galileo Galilei publishes *Discorsi e dimostrazioni matematiche intorno a due nuove scienze* in which he studied movement in pendulums and objects rolling down planes, and countered Aristotle's argument.	**Newton's Second law** says a change in motion is proportional to the force applied: *F=ma*.	Some claim that science cannot explain how a **bumblebee** can fly, but this is a fallacy.	Newton's laws are central to **engineering**.	We continue to fill in gaps, such as the understanding of how **kangaroos** can apparently use more energy in motion than they consume.
1687 Isaac Newton's masterpiece *Philosophiæ Naturalis Principia Mathematica* is published containing his laws of motion and mathematics of gravitational attraction.	**Newton's Third law** says that every action has an equal and opposite reaction.		Understanding how **natural things** work also requires Newton's laws.	
	Simple machines magnify force by reducing the distance over which the force is applied.			

● ● ○ ● ○

ELECTROMAGNETISM

THE **ESSENTIAL** IDEA

"ELECTRICITY IS OFTEN CALLED WONDERFUL, BEAUTIFUL; BUT IS
SO ONLY IN COMMON WITH THE OTHER FORCES OF NATURE."

MICHAEL FARADAY, 1858

Electromagnetism, combining the effects of electricity and magnetism, is arguably the most pervasive force of nature, though the effect that it has in everyday life is often not obvious. As astronauts have demonstrated, we can live without the much weaker force of gravity, but without electromagnetism we simply would not exist.

The more obvious aspect of electricity in wires and magnets is only the tip of the iceberg. It is electromagnetism that holds atoms together and that binds atoms into molecules. It is electromagnetism that allows us to sit on a chair. Electromagnetic effects are behind all the biological processes that make life possible – both in direct electrical effects and in chemistry, which is fundamentally electromagnetic.

To add to electromagnetism's contributions, light is also an electromagnetic effect. Photons of light carry the electromagnetic force between charged items, and the light from the Sun that enables us to see and that warms the Earth sufficiently to make life possible is pure electromagnetic energy.

 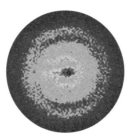 

Above left: The electrical currents in the bulb heats the filament to produce light.
Above middle: Atomic structure of a hydrogen atom, held together by the electromagnetic force.
Above right: The magnet's ability to attract metal is another example of the electromagnetic force in action.

ORIGINS

Static electricity was known to the ancients – the words "electron" and "electricity" comes from the Greek word for amber, because amber gets an electrical charge when it is rubbed. Similarly, electric fish were documented by the Ancient Egyptians more than 4,700 years ago. By the eighteenth century, electricity, with its ability to shock and pick up light objects, had become a parlour entertainment. Natural philosophers took a greater interest in its relation to nature, notably the American statesman Benjamin Franklin, who researched the subject widely in the 1740s, including his famous kite experiment with lightning (though it is thought he probably didn't undertake it himself).

MAGNETS

Interest was also growing in magnets, which had first been studied scientifically by English scientist William Gilbert in 1600. Initially there was some confusion between the two concepts, but in the nineteenth century they were identified as interacting parts of the overarching field of electromagnetism.

Key developments would include the Italian scientist Alessandro Volta's 1800 invention of the electrical battery, and work by Danish scientist Hans Ørsted and French scientist André-Marie Ampère before English scientist Michael Faraday pulled the concept together, inventing the electrical motor in 1821 and establishing the essential concept of electrical and magnetic fields – effectively the space of influence around an electromagnetic source – which had values that varied with time and position. Faraday's ideas would be built on by the Scottish physicist James Clerk Maxwell, who provided the mathematics to explain the interaction of electric and magnetic fields, and realized in 1864 that it should be possible to send an electromagnetic wave through the fields, which he calculated would travel at the speed of light.

"ELECTROMAGNETIC THEORY WAS TRANSFORMED BY QUANTUM PHYSICS, WITH THE REALIZATION THAT, THOUGH IT ACTED AS IF IT WERE A WAVE, LIGHT WAS COMPOSED OF A STREAM OF PARTICLES KNOWN AS PHOTONS."

Electromagnetic theory was transformed by quantum physics, with the realization that, though it acted as if it were a wave, light was composed of a stream of particles known as photons. The crowning glory in the understanding of electromagnetism through quantum theory would be quantum electrodynamics (QED), which won the Nobel Prize in Physics in 1965 for American physicists Richard Feynman and Julian Schwinger and Japanese physicist Sin-Itiro Tomonaga.

Top: The electric eel (strictly a fish, not an eel) provided one of the early natural examples of electricity.
Bottom: Michael Faraday (1791–1867)

KEY **THEORIES** AND **EVIDENCE**

EQUATIONS, WAVES AND PARTICLES

"GOD RUNS ELECTROMAGNETICS ON MONDAY, WEDNESDAY
AND FRIDAY BY THE WAVE THEORY, AND THE DEVIL RUNS IT BY
QUANTUM THEORY ON TUESDAY, THURSDAY AND SATURDAY."

WILLIAM LAWRENCE BRAGG, 1978

Unlike gravity, which is always attractive, electromagnetism can attract or repel. We refer to there being two, opposite electrical charges (positive and negative) or magnetic poles (north and south). The electromagnetic force produced between charged bodies is vastly more powerful that gravity. We think of gravity as being strong because we live in contact with a very large source of gravity – the Earth. But consider a fridge magnet. The whole massive Earth's gravitational pull is trying to get that magnet to fall off the fridge. All that is holding it up is the tiny magnet's electromagnetic force, but the electromagnetism wins.

Many of the electromagnetic effects we experience in everyday life are due to the electromagnetic nature of the atom. Each atom has a positively charged central nucleus along with one or more negatively charged electrons in the space around the nucleus, held in place by the electromagnetic force. Electrons also feature in most modern technology, as it is electrons flowing through wires that constitute current electricity, whether it's from our mains sockets or inside the processor of a computer.

Moving electricity produces magnetism, while moving magnets produce electricity. This is the basis behind electromagnets, electrical generators and motors. It also means that a changing level of electrical field will produce a changing level of magnetic field, which in turn produces

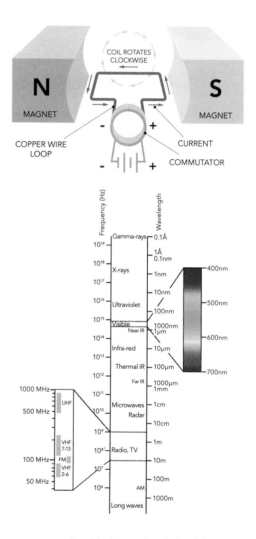

Top: The interaction of electricity and magnetism in an electric motor.

Bottom: The electromagnetic spectrum, including the narrow range of visible light.

Opposite: An electromagnetic wave showing changing levels of electrical (blue) and magnetic (grey) fields.

a changing electrical field (and so on), resulting in an electromagnetic wave. There is a wide spectrum of such waves from radio, microwaves and infrared through to ultraviolet, X-rays and gamma rays, with the most familiar being visible light, roughly located in the centre of the electromagnetic spectrum. Light is exactly the same thing as radio or X-rays, for example. The only way it differs is in the energy it carries; seen as a wave, this is reflected in its wavelength (shorter means higher energy) or frequency (higher means higher energy).

MAXWELL'S EQUATIONS

James Clerk Maxwell produced a series of equations that described exactly how electricity and magnetism interacted. His original formulation involved 20 separate formulae, but they were condensed down by the English physicist Oliver Heaviside into four very simple-looking equations that describe precisely how electromagnetism functions.

"THE NATURE OF LIGHT HAD BEEN THE SUBJECT OF SPECULATION FOR MILLENNIA."

The nature of light had been the subject of speculation for millennia. Although Isaac Newton had thought it to be made up of a flow of particles he called corpuscles, his contemporary Christiaan Huygens suggested that light progressed as a wave. It had been thought that this was proven in the early 1800s by English scientist Thomas Young, when he showed that light waves interfered with each other, as did waves in water. In interference, if waves are oscillating in the same direction when they cross, they magnify the wave, while waves vibrating in opposite directions cancel each other out. But the development of quantum theory made it clear that light was made up of a stream of particles called photons, even though quantum physics ensured that they produced wave-like behaviour.

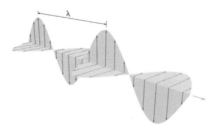

EVIDENCE

Nineteenth-century researchers undertook many experiments that demonstrated the relationship between electricity and magnetism, and Maxwell's mathematics made it possible to calculate the effects, and to predict the existence of an electromagnetic wave. Maxwell estimated this wave would travel at around 310,700 kilometres per second – the best estimate for the speed of light at the time was 314,850 kilometres per second, which surely could not be a coincidence.

The quantum nature of light was originally suggested by the way that so-called "black bodies", which absorb electromagnetic waves of all frequencies, produce unexpected spectra of light, but was given huge support by Einstein's 1905 explanation of the photoelectric effect that won him the 1921 Nobel Prize in Physics. Modern technology makes it possible to generate individual photons of light.

CRITICS

"FROM A LONG VIEW OF THE HISTORY OF MANKIND – SEEN FROM, SAY, TEN THOUSAND YEARS FROM NOW – THERE CAN BE LITTLE DOUBT THAT THE MOST SIGNIFICANT EVENT OF THE 19TH CENTURY WILL BE JUDGED AS MAXWELL'S DISCOVERY OF THE LAWS OF ELECTRODYNAMICS."

RICHARD FEYNMAN

In the eighteenth century there was considerable debate over the nature of the "electrical fluid" – the workings of electricity seemed quite similar to that of liquids in a pipe, as a result of which a number of electrical terms were derived from plumbing. For example, we speak of an electrical current, and the British term for a vacuum tube was a valve. Amusingly, when it was arbitrarily decided on which way a current flows, they guessed wrong – the actual flow of electrons is in the opposite direction. Some originally thought that positive and negative electrical charges were the result of two separate types of fluid.

When Maxwell came up with his equations, physics was still primarily an experimental science, and many leading physicists of the day, including William Thomson and Michael Faraday, could not understand Maxwell's work, which was based on purely mathematical considerations. But his approach would shape the future of physics, to the extent that Richard Feynman said: "From a long view of the history of mankind – seen from, say, ten thousand years from now – there can be little doubt that the most significant event of the 19th century will be judged as Maxwell's discovery of the laws of electrodynamics."

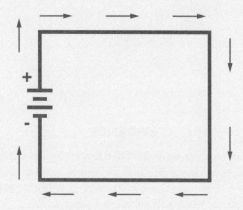

Above: By convention, electrical current flows from positive to negative, where electrons actually travel the opposite way.
Opposite top: Electricity is an essential driver of modern technology.
Opposite bottom: Electromagnetic waves carry energy from the Sun to heat and light the Earth.

WHY IT **MATTERS**

As one of the four fundamental forces of nature (the other three are gravity and the strong and weak nuclear forces), electromagnetism is central to our existence. Without it, matter could not interact with other matter other than through the very weak gravitational force (and matter would not exist in its current form as it is electromagnetism that holds atoms together).

Electromagnetism is also responsible for the light that both makes it possible for us to see and carries energy from the Sun, heating the Earth to a temperature that makes life possible. And chemical and electro-chemical electromagnetic processes are responsible for pretty much everything that happens within a living organism and gives it life.

However, the natural side of electromagnetism is only the start. After steam kicked off the industrial revolution it was electricity that carried it to new heights in the nineteenth century. Our modern lives are made possible by the electrical devices and the electronics that have flourished since the 1950s.

At one point, electricity and magnetism were considered so magical that it was assumed they would somehow provide medical cures – and some still cling to this in the use of magnetic bracelets and so forth. But what the Victorians could not have foreseen is the huge impact modern electrical and electronic technology would have on medicine, from diagnostic devices to X-ray machines and hi-tech scanners.

In every respect, we inhabit an electromagnetic world.

FUTURE **DEVELOPMENTS**

The basics of electricity and magnetism have been well established since the end of the nineteenth century, although we are still making discoveries around special materials, whether it is the atom-thick substance graphene, the best electrical conductor known, or the way that at extremely low temperatures, bodies expel magnetic fields, causing magnets to levitate above their surface.

"FOR MORE THAN 40 YEARS, COMPUTING HAS FOLLOWED 'MOORE'S LAW', WHICH STATES THAT THE NUMBER OF COMPONENTS (EFFECTIVELY THE POWER) OF A COMPUTER PROCESSOR DOUBLES EVERY TWO YEARS."

Developments in the field are mostly in electronics. For more than 40 years, computing has followed "Moore's law", which states that the number of components (effectively the power) of a computer processor doubles every two years. But we are reaching physical limits, meaning that new materials and approaches, such as quantum computers, need to be used.

We are seeing strong pressure to move from fossil fuels to electricity as a primary source of energy, which means that developments in electricity generation – particularly from solar cells and wind energy – continue to develop. We are also seeing electricity transmitted greater distances, prompting a move from the current high voltage alternating current (AC) distribution networks – which have been used for the relative ease of changing voltages using AC – to look at the potential for high voltage direct current (DC) distribution, which loses less energy in transmission.

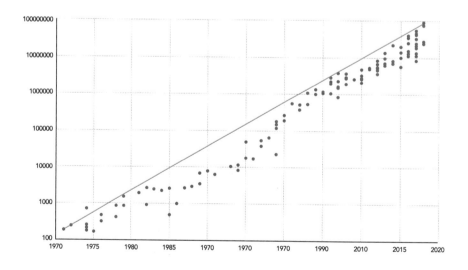

Above: Moore's Law, showing the (as yet) relentless rise in the power of computers.

THE **ESSENTIAL** SUMMARY

ORIGINS	KEY THEORIES AND EVIDENCE	CRITICS	WHY IT MATTERS	FUTURE DEVELOPMENTS
People have been aware of static electricity and magnetism for **thousands of years**. **1600** William Gilbert writes a tract on magnets and electricity called *De Magnete*. **1740s** Benjamin Franklin undertakes experiments, including his famous thunderstorm kite. **1800** Volta invents the battery. **Mid-19th century** Hans Ørsted, André-Marie Ampère, Michael Faraday and others establish the basics. **1860s** James Clerk Maxwell produces electromagnetic equations and theory of light.	**Electromagnetism is far stronger than gravity** and produces electrical charges/magnetic poles. **Atoms** are held together by electromagnetism, while moving electrons are current electricity. Interacting electrical and magnetic fields produce **electromagnetic waves** across a whole spectrum from radio to gamma rays, including **light**. Although light acts like a **wave**, it is composed of a stream of particles known as **photons**. Maxwell's equations enabled him to predict the speed of electromagnetic waves – which was the same as the **speed of light**. Einstein's explanation of the **photoelectric effect** was strong evidence for the existence of photons.	Eighteenth-century scientists argued over the number of **fluids** involved in electricity – and described **electrical current** flowing in the wrong direction. When Maxwell came up with his **mathematical description** of electromagnetism most physicists were **experimentalists** who couldn't understand the mathematics Maxwell used.	Electromagnetism is one of the **four fundamental forces** of nature. Electromagnetism enables **objects to interact** with each other and makes it possible for **matter to exist at all**. Electromagnetism is also behind **light**, without which we couldn't see and **heat reaches us from the Sun**, making life possible. Electromagnetism put the **industrial revolution** into a new gear: without **electrical and electronic devices**, modern life would be completely different.	New materials such as **graphene** are revolutionizing the development of new electronic technology. New approaches such as **quantum computing** are creating innovative possibilities for the future of electronics. The move from fossil fuels to electricity as our primary energy source is driving the development of **electricity generation**, particularly solar and wind, while long-range distribution through **high-voltage DC** is becoming feasible.

LIGHT

THE **ESSENTIAL** IDEA

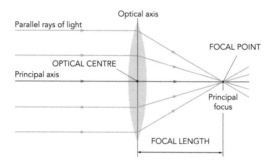

"FROM WHAT IT HAS BEEN SAID IT IS ALSO EVIDENT, THAT THE WHITENESS OF THE SUN'S LIGHT IS COMPOUNDED OF ALL THE COLOURS WHEREWITH THE SEVERAL SORTS OF RAYS WHEREOF THAT LIGHT CONSISTS, WHEN BY THEIR SEVERAL REFRANGIBILITIES THEY ARE SEPARATED FROM ONE ANOTHER, DO TINGE PAPER OR ANY OTHER WHITE BODY WHEREON THEY FALL."

ISAAC NEWTON, 1704

Light is essential for sight and carries energy from the Sun that keeps the Earth inhabitable. Optics, the study of the way that light can be manipulated, plays its part in everyday life in everything from eyeglasses to the lenses on smartphone cameras.

The colours of light that we see correspond to different energies in the photons. White light is a mix of the colours of the spectrum, which can be split out by a prism or raindrops to form a rainbow. When light colours mix, they produce different apparent colours because of the way that our eyes work.

ATOMIC ORIGINS

Atoms absorb and give off particular colours of light corresponding to the energy required to push an electron to a higher level in an atom. This means that when white light falls on an object, some colours are absorbed, some are re-emitted. The colours that aren't absorbed define the colour the object appears to be. For example, a red apple absorbs the colours of the visible spectrum except red.

The speed of light has a fixed value in any medium (299,792,458 metres per second – around 186,000 miles per second – in a vacuum): a key constant of nature.

ORIGINS

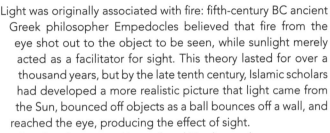

Light was originally associated with fire: fifth-century BC ancient Greek philosopher Empedocles believed that fire from the eye shot out to the object to be seen, while sunlight merely acted as a facilitator for sight. This theory lasted for over a thousand years, but by the late tenth century, Islamic scholars had developed a more realistic picture that light came from the Sun, bounced off objects as a ball bounces off a wall, and reached the eye, producing the effect of sight.

One key figure was Hasan Ibn al-Haytham, who wrote a text on optics around 1020. As well as correcting the model of light from the eye, he built on Greek mathematician Euclid's idea of beams of light travelling in straight lines to produce detailed ray diagrams of the progress of light when it struck flat and curved mirrors. Al-Haytham also studied refraction – how light bends when it moves from one medium to another, using it to estimate the thickness of the atmosphere. Optics continued to be studied over the following centuries in the new European universities, but the next major step forward came from Isaac Newton. His first work was inspired by using a prism, reporting in the early 1670s that the colours in a rainbow were all present in white light, rather than being produced by the prism, as was thought at the time.

"LIGHT WAS ORIGINALLY ASSOCIATED WITH FIRE: FIFTH-CENTURY BC ANCIENT GREEK PHILOSOPHER EMPEDOCLES BELIEVED THAT FIRE FROM THE EYE SHOT OUT TO THE OBJECT TO BE SEEN, WHILE SUNLIGHT MERELY ACTED AS A FACILITATOR FOR SIGHT."

SPEED OF LIGHT

The speed of light was debated for centuries – some thought it travelled instantly from place to place, while others thought its velocity was finite. However, in 1676 Danish astronomer Ole Rømer accidentally discovered that light had a measurable speed when tracking the moons of Jupiter. Light's speed based on Rømer's measurements was around 220,000 kilometres per second (137,000 miles per second), but would be refined to around 300,000 kilometres per second (186,000 miles per second).

The nature of light was also long disputed, with Newton thinking it a stream of particles – as it could pass through empty space – and others, such as Dutch scientist Christiaan Huygens, thinking it was a wave, which required there to be some kind of material, known as the ether, to carry it. As we have seen, it was in the 1860s that Maxwell established light as an interaction of electricity and magnetism.

Opposite: Optics involves the study of the manipulation of light by lenses.
Top: Empedocles (c490–430 BCE).
Bottom: Changes in timing of the movements of Jupiter's moons provided the first measurement of light speed.

KEY **THEORIES** AND **EVIDENCE**

ELECTROMAGNETIC RADIATION, COLOURS AND SPEEDS

"I WANT TO EMPHASIZE THAT LIGHT COMES IN THIS FORM – PARTICLES. IT IS VERY IMPORTANT TO KNOW THAT LIGHT BEHAVES LIKE PARTICLES, ESPECIALLY FOR THOSE OF YOU WHO HAVE GONE TO SCHOOL, WHERE YOU WERE PROBABLY TOLD ABOUT LIGHT BEHAVING LIKE WAVES. I'M TELLING YOU THE WAY IT DOES BEHAVE – LIKE PARTICLES."

RICHARD FEYNMAN, 1985

As we have seen, light is electromagnetic radiation in the form of a stream of photons, although because of its quantum nature it can behave like a wave. The rainbow spectrum of colours that we can see form a small section near the centre of the full electromagnetic spectrum. White light from the Sun can be split into the rainbow of colours when it passes through a suitable shaped material because of refraction. When light passes from one material to another it bends, but the angle of bending is different depending on the colour, meaning that the colours split out.

PERCEIVING COLOURS

The way that we perceive colours of light is dependent on the mechanism of the eye. Our eyes have three types of colour-sensitive cell, each covering a range of light energies (wavelengths). The corresponding three primary colours of light are red, green and blue. These can be combined at different levels to produce any perceived colour, some of which (such as magenta or brown) are not in the spectrum.

When light hits an object, individual photons of light are absorbed by atoms. The energy of the photon pushes one of the outer electrons of the atom to a higher energy level. Some of these electrons will then drop back down, re-emitting a photon. It is the colours that are re-emitted that establish what we see as the colour of that object. Because these colours are the result of subtracting some of the light, pigments combine to produce colours based on the secondary colours of magenta, cyan and yellow. These are the colours used in printing, which are approximated to in the incorrect "primary" paint colours often taught in school of red, blue and yellow.

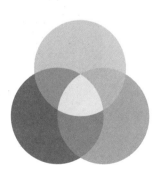

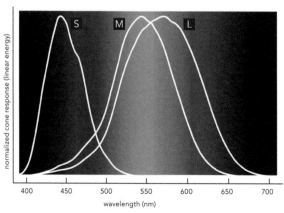

Lenses, which we use to focus light, are specially shaped so that the refraction that occurs when light passes from air to glass and then back from glass to air, bends beams of light by appropriate amounts to bring them together at a focal point. Simple blocks of glass bend different colours to different degrees, but modern lenses are usually specially shaped to avoid this distorting effect.

REFRACTION

The reason that refraction occurs is that light travels at different speeds in different materials, following what's sometimes called the "Baywatch principle". When a lifeguard heads for a drowning person, they don't go in a straight line but go further on the beach, where they can travel faster, then change angle into the water; similarly, light takes the route that minimizes journey time.

Light travels fastest in a vacuum – its speed there is exactly 299,792,458 metres per second, and the definition of a metre is now the distance light travels in 1/299,792,458th of a second. This is a maximum speed limit for the universe – nothing can travel faster than light. When we look up at the stars, we see them as they were in the past because light travels at a finite speed. Light has been travelling for over 13 billion years from the most distant galaxy that has been observed.

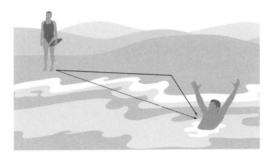

EVIDENCE

Newton's experiment that proved the rainbow colours were in white light rather than being produced by the prism involved sending a single colour from a prism through a second prism, where it emerged unchanged, not affected by the assumed impurities in the glass. The primary colours were demonstrated to be red, green and blue by Scottish scientist James Clerk Maxwell using a "colour wheel" – a spinning disk on which he put different coloured papers, so that the colours were combined in the eye. After Rømer, the speed of light was measured by sending through high-speed spinning gear wheels and mirrors, but now it is determined using interference patterns between laser beams.

Opposite left: The primary colours of light: red, green and blue, combine to form visible colours.
Opposite right: The three types of cone in the human eye respond to different parts of the colour spectrum.
Top: The Baywatch principle: it is quicker to take a longer route if more can be traversed at high speed.
Bottom: James Clerk Maxwell (1831–1879)

CRITICS

Remarkably, the idea that the primary colours were red, green and blue was still disputed in the second half of the nineteenth century. In 1870, 15 years after Maxwell published his result, non-scientists were still arguing for red, yellow and blue.

There was much debate over the speed of light before it became possible to measure it. Clearly it was incredibly fast. Thunder and lightning gave a clear example of this – we hear the sound of the thunder seconds later than we see the flash of the lightning. French philosopher René Descartes, for example, was amongst those who were convinced that light got from A to B instantaneously. He thought it effectively a pressure on the eyeball, transmitted through an invisible ether that filled all of space. He imagined it was like pushing one end of a rigid object and the other end moves instantly (though in reality a push passes down a physical object in a wave, not instantly).

> "FRENCH PHILOSOPHER RENÉ DESCARTES, FOR EXAMPLE, WAS AMONGST THOSE WHO WERE CONVINCED THAT LIGHT GOT FROM A TO B INSTANTANEOUSLY."

There have always been those who doubt the light-speed limit, which arises from Einstein's special theory of relativity. They point out, for example, that the end of a light beam that is sweeping around, like that of a lighthouse, can move faster than the speed of light, and that cosmologists believe that the universe has expanded faster than the speed of light. However, the limit is only for a physical object moving through space. This is the hope of those who would like to build a "warp drive" – that there is some way to avoid the light-speed barrier by warping space around a ship.

Above: The end of a lighthouse beam can sweep around faster than light speed.

WHY IT **MATTERS**

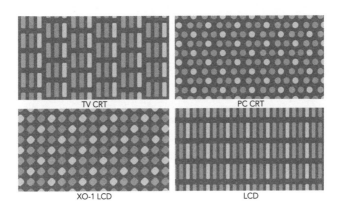

| TV CRT | PC CRT |
| XO-1 LCD | LCD |

"ONE OF THE REASONS THAT OPTICS WAS AN EARLY DEVELOPMENT IS ITS VALUE IN CORRECTING EYESIGHT – EYEGLASSES DATE BACK TO AROUND THE THIRTEENTH CENTURY. "

As we have seen with electromagnetism in general, it is light that both enables us to see and carries energy from the Sun to warm our planet to a habitable temperature. An understanding of how different colours combine in the eye was of interest early on to understand colour blindness, which usually involves one or more of the light-sensitive cell types in the eye malfunctioning. However, in modern times it has added importance.

Whenever we use a device with a colour screen – a TV, a computer, a phone – that screen works using the principle that Maxwell demonstrated of using different combinations of red, blue and green to produce the whole range of colours. Similarly, a modern colour printer requires a knowledge of the combination effects of the secondary pigment colours, magenta, cyan and yellow (plus black).

One of the reasons that optics was an early development is its value in correcting eyesight – eyeglasses date back to around the thirteenth century. Optical devices using both lenses and mirrors have also been helping us get a better view of the world and universe around us with the development of microscopes and telescopes from the seventeenth century onwards. And it is light (along with other energies of electromagnetic radiation) that enables us to explore the universe and look back into its past, thanks to light's finite speed.

Above: Red, green and blue pixels combine in screens to produce the full range of colours.

FUTURE **DEVELOPMENTS**

Our ability to produce light has taken many recent steps forward, notably the move from incandescent bulbs to light-emitting diodes (LEDs). We are also still scratching the surface of the potential of a related type of light source, the laser. The first example was constructed in 1960, based on a theory of Albert Einstein's.

Lasers are very different from a traditional light source like the Sun, a candle or a lightbulb, because the light they produce is coherent. This means that all the photons they produce have the same energy and are in phase. Thinking of the light as waves, the waves have the same wavelength and move in step. This results in light that can travel long distances without dissipating and is capable of concentrating energy for specialist cutting processes.

> "SPECIAL LENSES CAN BE PRODUCED THAT ARE CAPABLE OF MAGNIFICATIONS THAT ARE PHYSICALLY IMPOSSIBLE WITH AN ORDINARY LENS"

Most current optical systems rely on components that were familiar hundreds of years ago: lenses and mirrors. The effects of all optical devices are quantum, but a new breed of purely quantum devices is available, which can manipulate light in ways that would not otherwise be possible. For example, special lenses can be produced that are capable of magnifications that are physically impossible with an ordinary lens, and special materials can bend light around objects, forming a small-scale "cloaking" device.

Above: Lasers used in the display can travel long distances without dissipating.

THE **ESSENTIAL** SUMMARY

ORIGINS	KEY THEORIES AND EVIDENCE	CRITICS	WHY IT MATTERS	FUTURE DEVELOPMENTS
5th century BC – Empedocles believes that light is fire from our eyes, enabling us to see. **1020** Hasan Ibn al-Haytham uses a correct model of light from a source bouncing off objects and reaching the eye, produces ray diagrams for mirrors and uses refraction to estimate the depth of the atmosphere. **1670s** Newton shows that the colours of the rainbow are contained within white light. **1676** Ole Rømer measures the speed of light (unintentionally). **1690** Christiaan Huygens puts forward his wave theory of light. **1860s** Maxwell produces electromagnetic theory of light.	**Light is a stream of photons** which, because of their quantum nature, can act like a wave. **White light contains all the colours of the rainbow**, which can be split out with a prism as different colours are refracted to different angles. The **primary colours** of light, from which all others can be made, are **red, green and blue**. Light hitting an object is partly absorbed, with the energies re-emitted defining its colour. This results in the **secondary pigment colours** of **magenta, cyan and yellow**. Lenses bend light due to their shape and focus at a point. This effect is produced by **refraction**, resulting from light travelling at different speeds in different materials. Light's speed in a vacuum, **299,792,458 metres per second**, is the fastest anything can travel through space.	Many non-scientists **refused to accept** red, green and blue as the primary colours in the nineteenth century. Descartes and others thought that **light travelled instantly**, like pushing one end of a stick and getting an instant movement in the other. Many have tried to find ways around the **light-speed limit**, mostly based on warping space, as the limit is only for passing through space.	Light enables us to **see** and carries **energy from the Sun** to warm the Earth. Any device with a **colour screen** makes use of mixing the three primary colours: red, green and blue. Optics has had a long role in **correcting eyesight** using glasses. Optics using lenses and mirrors has been central to the development of astronomy, microscopy and long-distance viewing in the **telescope** and **microscope**. Light's speed limit gives us the opportunity to examine the **prehistory of the universe**, as the further away we look, the further back in time we see.	We are constantly improving light sources; for example, in the move from incandescent bulbs to **light-emitting diodes** (LEDs). The **laser** has provided us with a whole new type of light source, for which new uses are still being found. New **quantum optics** can bend light in otherwise impossible ways to create **super-lenses** and **cloaking devices**.

QUANTUM THEORY

THE **ESSENTIAL** IDEA

"THE THEORY OF QUANTUM ELECTRODYNAMICS DESCRIBES NATURE AS ABSURD FROM THE POINT OF VIEW OF COMMON SENSE. AND IT AGREES FULLY WITH EXPERIMENT. SO I HOPE YOU CAN ACCEPT NATURE AS SHE IS – ABSURD."

RICHARD FEYNMAN, 1985

Quantum theory describes how the universe works on the scale of atoms, sub-atomic particles and photons. Most of the world around us consists of these particles, yet quantum physics seems counter-intuitive because, on this scale, reality is fundamentally different from our day-to-day experience. Quantum theory depends on three straightforward concepts:

Quanta: the fundamentals of reality are not continuously variable but come in chunks, known as quanta. The consequences of this apparently small change in our understanding were huge.

Probability: prior to the quantum revolution it was assumed that if we had perfect information, we could predict the future exactly. But quantum physics makes it clear that for quantum particles the future is a matter of probabilities. These probabilities are calculable, but the outcome can never be sure until it happens.

Superposition: because the state of a quantum particle is driven by probability, a particle is in every possible state before it is observed, each with a specific probability, a situation known as superposition.

ORIGINS

In 1900, German physicist Max Planck was attempting to explain the "ultraviolet catastrophe". Matter gives off electromagnetic radiation that varies with temperature – so, for example, as we heat up a piece of metal it glows red, then yellow, then white. Theory predicted that at room temperature objects should give off intense amounts of ultraviolet light. Planck made theory match the actual colours observed by pretending that light came in little packets called "quanta". He did not believe they existed but regarded them as useful for calculation.

ALBERT EINSTEIN

Five years later, a younger German physicist, Albert Einstein, working as a patent clerk in Switzerland, explained another odd phenomenon by assuming quanta were real. When light is shone on some metals, an electric current starts to flow in them, as energy from the light knocks electrons out of the metal atoms. If light were a continuous wave, as was thought at that time, you would expect that any colour of light could make this happen if the light was bright enough. But instead, only high-energy light, such as blue light, did. Einstein realized this suggested that light really did come in Planck's quanta, rather than as a continuous wave. Light quanta would later be called photons.

NIELS BOHR

Einstein's theory was used in 1913 by a young Danish physicist, Niels Bohr, to explain how different atoms gave off particular colours of light when they were heated (and absorbed the same colours from passing light). Bohr had been trying to understand the structure of atoms – he had considered the possibility that they were like miniature solar systems with electrons orbiting the heavy central nucleus, as postulated by New Zealand physicist Ernest Rutherford. However, such an atom would collapse, as orbiting electrons would give off light and spiral into the nucleus. Bohr imagined the electrons on fixed paths around the atoms with the amount of energy they could receive or give off quantized rather than continuous. Then they could only undergo fixed sizes of jumps, known as a quantum leaps, and give off specific colours of photons, exactly as found experimentally.

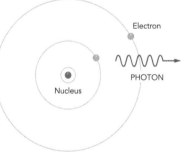

These three developments provided the foundation of a theory that would be developed in the 1920s to provide an effective but mind-boggling explanation of the world of the very small.

Opposite: For quantum particles the future is a matter of probabilities.
Top: Max Planck (1858–1947).
Middle: Albert Einstein (1879–1955).
Bottom: Atoms give off energy in the form of photons of light when electrons drop down in energy level.

KEY **THEORIES** AND **EVIDENCE**

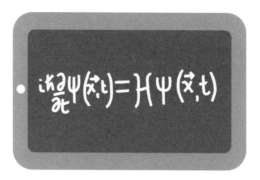

$$i\hbar \frac{\partial}{\partial t}\psi(\vec{x},t) = H\psi(\vec{x},t)$$

QUANTUM MECHANICS, QED AND INTERPRETATIONS

After the initial foundations, quantum theory was expanded. At the heart of the new theory was Erwin Schrödinger's equation, describing how the state of a quantum system evolves over time. A quantum system is one or more quantum particles, while the state describes all the properties of those particles, such as their location, momentum, spin and more. According to the theory, rather than these properties having exact values, the quantum system is in a superposition of all possible values, each with a probability described by Schrödinger's equation.

UNCERTAINTY PRINCIPLE

One major implication of quantum theory is the uncertainty principle. This states that there are pairs of properties of quantum systems that are inextricably linked. One pair is position and momentum. The principle tells us that the more accurately we know one of the pair, the less certain we can be about the other. So, for example, if we knew exactly where a particle was, we would be totally unsure about its momentum. Another important pairing is energy and time. If we take a very narrow window of time, the amount of energy in a quantum system – even empty space – is hugely variable. Because energy and matter are interchangeable, the implication is that new particles can pop in and out of existence.

ENTANGLEMENT AND QED

Another important outcome of quantum theory is entanglement. This tells us that it is possible to produce two or more quantum particles in a special state where they are part of the same system, even if separated at any distance. A change to one particle will be instantly reflected in the other(s), no matter how far apart they are.

Above: Schrödinger's equation describes how a quantum system evolves over time.
Opposite top: Quantum physics explains why a mirror reflects at unexpected angles when strips are missing.
Opposite bottom: The many worlds interpretation suggests the universe splits each time there are two possible outcomes of a quantum interaction.

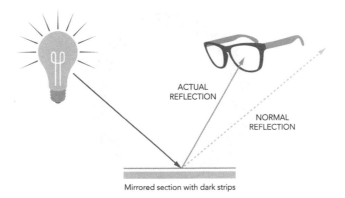

ACTUAL
REFLECTION

NORMAL
REFLECTION

Mirrored section with dark strips

Perhaps the most important aspect of quantum physics is QED (quantum electrodynamics). This describes how quantum particles interact. QED showed that apparently simple interactions, such as reflection in a mirror, are deceptive. When light reflects off a mirror, it doesn't do so at a neat reflected angle. Instead, there are probabilities of the photons taking every possible path, including reflecting at crazy angles – but usually the unexpected paths cancel each other out. However, removing parts of the mirror to omit some cancelled routes results in a reflection at an unexpected angle.

Quantum theory is unique in having several interpretations. What actually happens is clearly predicted by the mathematics – but not *why*. Many physicists stick to a basic interpretation known as the Copenhagen interpretation. This says we can't know what's going on until we measure things – all that exists are probabilities. This is typified as "shut up and calculate".

"WHAT ACTUALLY HAPPENS IS CLEARLY PREDICTED BY THE MATHEMATICS – BUT NOT *WHY*."

Others want something describable beneath the quantum phenomena – but providing an explanation produces its own weirdness. So, for example, one interpretation, "many worlds", suggests that every time a quantum particle interacts with the world around it, the universe splits so that all possible outcomes occur – but we only experience one path through these splits. Another interpretation, the Bohm interpretation, suggests that each particle has an associated "pilot" wave that guides it. Here, the probabilities reflect an underlying (but inaccessible) reality – but the cost of this is that every particle in the universe must be constantly linked to every other.

EVIDENCE

Since the 1920s, many experiments have confirmed the theory; every modern electronic device adds to the evidence, as their design is based on quantum physics. An early experiment backing up the theory was the double slit, devised in the 1800s to prove that light was a wave. Light is passed through two narrow slits and shone onto a distant screen. Rather than produce two bright regions, the result is a series of dark and light fringes. This was explained by a process known as interference, a wave effect from two waves interacting.

Experiments showed that the fringes still build up if individual quantum particles – electrons or photons, for example – are sent through one at a time. If particles had a distinct location, they would pass through one slit and could not produce interference. But instead, as only waves of probability exist, these pass through both slits, interfering to produce fringes.

Other experiments show the ability of quantum particles to pass through barriers. As quantum particles do not have a distinct location, there is a possibility that a particle near a barrier could be on the other side without passing through it. This is known as quantum tunnelling. Not only can this be demonstrated (it's how flash memory works), but without such tunnelling, the Sun could not function. We are alive because of it.

"THIS IS KNOWN AS QUANTUM TUNNELLING ...
WE ARE ALIVE BECAUSE OF IT."

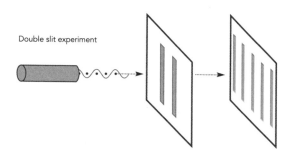

Double slit experiment

Top: The Sun would not function, generating energy that keeps us alive, without quantum tunnelling.
Bottom: The pattern of fringes in the double slit builds up even if particles are sent through one at a time.

CRITICS

Though a founder of quantum physics, Einstein became its harshest critic. He repeatedly hunted for flaws in the theory, and was particularly unhappy with its basis on probability, hence his remarks about God not playing dice. He wrote to a friend, "In that case, I would rather be a cobbler, or even an employee in a gaming house, than a physicist."

Another opponent of probabilistic outcomes was Erwin Schrödinger. He came up with his cat thought experiment to demonstrate the weirdness of quantum theory. The time it takes a radioactive particle to decay is governed by probability. Schrödinger imagined putting a cat in a closed box with a radioactive particle, a detector and a vial of cyanide that would be released when the particle decayed. The box is closed. After some time has elapsed, the particle is in a superposition state of decayed and not decayed. Is the cat both dead and alive until we open the box?

In reality, the interaction between the particle and the detector results in "decoherence" where the particle loses its superposition and takes on one state or another – and all the experimental evidence supports quantum theory working – but Schrödinger's cat remains entertaining.

WHY IT **MATTERS**

Quantum physics enables us to predict the behaviour of quantum particles with remarkable accuracy. It works wonderfully. As quantum physicist Richard Feynman observed, it is the most accurate theory in existence, enabling us to match reality with accuracy that is the equivalent of predicting the distance from New York to Los Angeles to the accuracy of the width of a human hair.

Between them, quantum theory and the general theory of relativity transformed our understanding of how the universe works as a whole, from the very small to the very large. These theories replace most earlier theories of physics (still taught in school) with approaches that are far closer to what is actually observed.

"GENERAL RELATIVITY IS OF LITTLE PRACTICAL USE. QUANTUM THEORY, THOUGH, IS CENTRAL TO MUCH OF OUR MODERN TECHNOLOGY."

However, there is a big difference between general relativity and quantum theory. General relativity is of little practical use. Quantum theory, though, is central to much of our modern technology. It has been estimated that 35 per cent of GDP in developed countries depends on products based on quantum theory. All electronics and a lot of modern optics could not be developed without an understanding of the quantum world. Modern devices such as lasers and MRI scanners are inherently quantum in nature. We live in a quantum age.

FUTURE **DEVELOPMENTS**

Although we have so many electronic devices with quantum physics at their hearts, there is much more to come. Quantum computers are being developed, which, instead of using bits with the value 0 or 1, use "qubits" – quantum bits, which can hold any value between 0 and 1 and can make multiple calculations simultaneously, enabling a fully functioning quantum computer to undertake calculations that would be impossible in any practical timescale with a conventional computer.

"QUANTUM ENTANGLEMENT IS BEING USED TO PRODUCE UNBREAKABLE ENCRYPTION"

Quantum entanglement is being used to produce unbreakable encryption and to produce a small-scale equivalent of a Star Trek transporter in quantum teleportation. And quantum optics, making use of new ways to manipulate photons, shows the way to produce impossibly powerful lenses and small-scale invisibility cloaks.

On the theoretical level, there is a fundamental disconnect between quantum theory and general relativity, the theory of gravity. The two are incompatible: general relativity assumes that the universe is continuous, where quantum theory implies it is granular. Ever since Einstein's time, physicists have looked for ways to bring together quantum physics and general relativity. The best supported approach, string theory, has significant problems and is losing support. It may be that the two can never be resolved – or that a totally new theory will be required to bring everything together under the same "theory of everything".

Above: General relativity is highly effective on the scale of galaxies, but as yet is incompatible with quantum theory.

THE **ESSENTIAL** SUMMARY

ORIGINS	KEY THEORIES AND EVIDENCE	CRITICS	WHY IT MATTERS	FUTURE DEVELOPMENTS
1900 Max Planck uses the idea that light energy was made of individual units called quanta to fix the ultraviolet catastrophe. **1905** Albert Einstein explains the photoelectric effect by assuming that Planck's imaginary quanta are real. **1913** Niels Bohr publishes his papers on the quantum structure of the atom. **1926** Erwin Schrödinger publishes a paper incorporating his equation describing the probabilistic nature of quantum properties. **1928** Werner Heisenberg introduces his uncertainty principle, relating pairs of quantum properties.	**Quanta:** physical phenomena come in chunks or "quanta". **Schrödinger equation:** the properties of quantum particles are only described by probability waves until they are measured. **Uncertainty principle**: pairs of quantum properties, such as position and momentum, are linked, so the more accurately one is measured, the less accurately the other can be known. **Entanglement**: quantum particles can be linked in such a way that a change in one is reflected instantly in the other, however far apart they are separated. **QED**: when light and matter interact, we must consider every possible path with different probabilities, producing unexpected results. **Interpretations**: the most common understanding is that only probabilities exist for the values of quantum properties until a measurement is made, when a specific value is discovered. This Copenhagen interpretation has a number of rivals including "many worlds", which tells us the universe splits every time quantum particles interact, and the Bohm interpretation, in which particles are guided by a special wave. **Evidence:** many experiments demonstrate the effectiveness of quantum theory in describing nature. All modern electronics depends on our understanding of the quantum.	**Albert Einstein**, one of the architects of quantum theory, was never convinced by its "spooky" nature. He felt there had to be hidden information rather than just probabilities behind quantum properties. **Erwin Schrödinger**, one of the top second-generation quantum physicists, was never comfortable with the Copenhagen interpretation and came up with his famous cat thought experiment to show why.	It is the most **fundamental theory** in physics, explaining how the basic building blocks of light and matter act and interact. It suggests that all events in nature are **probabilistic**. Our universe is nothing like the clockwork machine envisaged by classical mechanics. All modern **electronic technology** is based on quantum theory. Some modern devices such as lasers and MRI scanners could not even be conceived of without quantum physics.	The two main theories of physics, the general theory of relativity and quantum theory, are **fundamentally at odds**. This is a crucial scientific problem. Although theories such as **string theory** attempt to combine general relativity and quantum physics, none is yet satisfactory. Some believe one or both will eventually have to be replaced to have a successful "theory of everything". New quantum technologies such as **quantum encryption**, **quantum computing** and **quantum optics** are likely to produce significant breakthroughs in the future.

RELATIVITY: TIME AND SPACE

THE **ESSENTIAL** IDEA

"ALTHOUGH TIME, SPACE, PLACE AND MOTION ARE VERY FAMILIAR TO EVERYONE, IT MUST BE NOTED THAT THESE QUANTITIES ARE POPULARLY CONCEIVED SOLELY WITH REFERENCE TO THE OBJECTS OF SENSE PERCEPTION."

ISAAC NEWTON, 1687

Basic relativity is common sense. If we say something is moving, we have to ask: "With respect to what?" If you are sitting in an armchair reading this, you are not moving – with respect to the chair. But along with the Earth, you are moving at around 30 kilometres per second (18.6 miles per second) with respect to the Sun.

Relativity tells us that, for example, if we are in a steadily moving vehicle, we can't detect that motion without looking at the outside world. It also means that if two cars head towards each other, each travelling at 50 miles per hour (with respect to the road), their relative speed is 100 mph.

"EINSTEIN SUGGESTED THAT NO MATTER HOW YOU MOVE WITH RESPECT TO A BEAM OF LIGHT, IT ALWAYS TRAVELS AT THE SAME VELOCITY. "

Einstein's special theory of relativity goes a step further. Because light can only exist at a particular speed in any medium, Einstein suggested that no matter how you move with respect to a beam of light, it always travels at the same velocity. From this simple observation, time and space become entwined. Moving through space causes time to slow down, mass to increase and length to reduce. Einstein also deduced that $E=mc^2$, relating energy and mass (c is the speed of light).

ORIGINS

Basic relativity is often called Galilean relativity because it was Galileo Galilei who first specified the inability to detect motion within an enclosed boat moving at a steady speed, and discussed relativity in his *Discorsi e dimostrazioni matematiche intorno a due nuove scienze*. Galileo is said to have shocked a friend while being rowed at high speed across Lake Piediluco by throwing his friend's house key high in the air. The friend nearly dived in the water, assuming the key would be left behind by the fast-moving boat, but in fact it dropped straight back into Galileo's lap. Although the boat was moving with respect to the water, it was not moving with respect to the key.

"GALILEO IS SAID TO HAVE SHOCKED A FRIEND WHILE BEING ROWED AT HIGH SPEED ACROSS LAKE PIEDILUCO BY THROWING HIS FRIEND'S HOUSE KEY HIGH IN THE AIR. "

SPECIAL RELATIVITY

Albert Einstein's special theory of relativity was one of four major papers he published in 1905 while working as a clerk in the Swiss patent office. There had already been some work done on the topic by others, such as Dutch physicist Hendrik Lorentz and Irish physicist George FitzGerald, but Einstein pulled the whole concept together in a remarkable paper.

Although relativity is often presented as being complex and hard to understand, this is primarily down to Einstein's second piece of work on the topic, the general theory of relativity, which will come up under the subject of gravity. In reality, all that is needed to understand the special theory is high-school mathematics, though the implications are mind-boggling.

A few months after publishing his special theory, Einstein wrote a second paper, only a few pages long, which mathematically derived from the special theory the equation relating energy and matter, $E=mc^2$. At the time he thought it little more than an interesting oddity, but it would have major implications both in the workings of nuclear weapons and in understanding the early evolution of the universe.

Opposite: The special theory of relativity means that the passage of time is relative, not absolute.
Above: In a small extension to the special theory, Einstein established the equation that would make nuclear weapons possible.

KEY **THEORIES** AND **EVIDENCE**

RELATIVE MOTION, THE SPEED OF LIGHT AND SPACETIME

"THE THEORY OF RELATIVITY CONFERS AN ABSOLUTE MEANING ON A MAGNITUDE WHICH IN CLASSICAL THEORY HAS ONLY A RELATIVE SIGNIFICANCE: THE SPEED OF LIGHT."

MAX PLANCK, 1947

One of the reasons that it took so long to accept that the Earth was not stationary at the centre of the universe was that common sense suggests we would feel the effects of a moving Earth. Galilean relativity is logical, but it doesn't occur to us immediately because most of the motion we experience is with respect to the Earth: it can feel like this movement is in some sense absolute.

RELATIVE MOTION

However, relative motion is the reality we face. Imagine you were entirely alone in the universe – there was nothing else there. Are you moving? It would be impossible to say, because there would be nothing to measure your movement against. There is no measure – no absolute "frame of reference" as physicists say – against which that motion can be judged. We're just used to the Earth being so big and obvious that it acts as a useful local frame of reference.

Einstein took things further by noting that Maxwell had showed that light is an interaction between electrical and magnetic fields that can only occur at a certain velocity. Usually, if we travel alongside a moving object, as far as we are concerned that thing isn't moving. But if this were true of light, it would cease to exist whenever you moved. So, Einstein assumed that light always travels at the same speed.

Using simple examples, such as a light clock, it is possible to show that this effect of light means that time and space cannot be considered absolute. A light clock is a beam of light bouncing up and down between a pair of mirrors. If we imagine

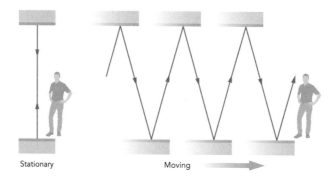

Stationary

Moving

Above: Seen from outside the moving spaceship, the light beam travels further than from the stationary position within the ship.
Opposite: Muon particles in cosmic rays travel so fast that time slows noticeably for them.

looking at a light clock on a spaceship travelling at high speed away from us, we would see that the light travelled at a diagonal to get from one mirror to the other. We see it travelling further than people on the ship would see. As light always goes at the same speed, the only way this is possible is if time is running slower on the ship, from our viewpoint – though the people on the ship would not notice it.

> "USING SIMPLE EXAMPLES, SUCH AS A LIGHT CLOCK, IT IS POSSIBLE TO SHOW THAT THIS EFFECT OF LIGHT MEANS THAT TIME AND SPACE CANNOT BE CONSIDERED ABSOLUTE. "

Similarly, the ship would, from our viewpoint, be compressed in the direction of travel and would have an increase in mass. As the speed of the ship approaches the speed of light, the mass would approach infinity. So it can be deduced that nothing with mass can reach the speed of light. The theory also shows that the concept of simultaneous events at locations separated in space is meaningless, as the relative timing changes if the observer is moving. The special theory moves away from separate concepts of space and time, providing a unified concept known as spacetime.

Einstein's extension to his original paper showed that the kinetic energy of a body moving at constant speed will decrease if it gives off light. As kinetic energy is dependent just on mass and velocity, the implication was a loss of mass, which he calculated as E/c^2 – so, although the equation never appears in the paper, the implication was that $E=mc^2$.

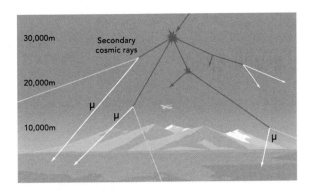

EVIDENCE

Galilean relativity is easily shown using simple equipment – Galileo himself suggested setting up experiments inside a steadily moving boat with no windows, where no movement can be detected. It also follows when we think about the relative motion of the Earth to other parts of the universe.

The special theory was derived from theoretical considerations, but there is no escaping the conclusions. However, since 1905, it has been tested many times and demonstrated to be true. For example, unstable particles called muons entering the Earth's atmosphere from space at a fair proportion of the speed of light exist significantly longer than they should before decaying, as their time is running slower seen from the Earth's surface.

CRITICS

There was little scientific resistance to Galilean relativity, however Newton insisted that there was absolute space and time. His absolute time was astronomical time, while absolute space was an imagined fixed framework of the universe with religious connotations. Newton presented an example using a rotating bucket of water that seemed to imply absolute space. The water "recede[s] from the middle and rise[s] up the sides" in Newton's words, staying like this even when it has caught up with the bucket and is not rotating with respect to it. Newton thought the rotation was with respect to absolute space.

"THERE WAS SURPRISINGLY LITTLE CRITICISM OF EINSTEIN'S SPECIAL THEORY OF RELATIVITY"

Of course, the Earth has an influence on it, but Newton had a point. After all, if something rotates in space, things inside it feel a force pulling them to the outside – it has been suggested as means to produce artificial gravity. But what is the spaceship rotating with respect to? All the evidence from experiments is that spacetime is relative – but this remains something of a puzzle, with some positing that the rotation is with respect to the whole universe.

There was surprisingly little criticism of Einstein's special theory of relativity, though the significance of his work on relativity was not widely recognized until he added the general theory in 1917 – it was for his work on quantum theory that he won the Nobel Prize in Physics.

WHY IT **MATTERS**

"THE BASICS OF THE SPECIAL THEORY OF RELATIVITY ONLY BECOME SIGNIFICANT WHEN OBJECTS ARE MOVING VERY QUICKLY WITH RESPECT TO EACH OTHER"

Galilean relativity is very important whenever we consider the interaction of moving bodies. An obvious example is a car crash, where it is relative speeds that are significant. Similarly, when flying in a plane there are two speeds to consider. Air speed – the speed with respect to the air is the speed that the plane flies at as far as its stability in the air is concerned. This is why aircraft tend to take off into the wind, as the result is that the plane moves faster through the air, enabling it to take off at lower speed compared to the ground. Ground speed, though, is essential to know when a plane will arrive at its destination.

The basics of the special theory of relativity only become significant when objects are moving very quickly with respect to each other – until then, Newton's laws work fine. However, it is important to understand fast-moving particles, whether in space or particle accelerators. The most significant aspect of special relativity for the everyday is $E=mc^2$, with practical implications for nuclear physics and theoretical implications for understanding the interplay of energy and matter. It is also the case that the special theory was the stepping-off point for the general theory of relativity, which took in acceleration and gave us our understanding of gravity.

Opposite: Future spaceships may have a rotating section to produce artificial gravity.
Above: The relative speed in a head-on collision is the sum of the two individual speeds.

FUTURE **DEVELOPMENTS**

The most positive potential application of $E=mc^2$ is in fusion reactors. These nuclear reactors use the same process as the Sun – fusing together lighter elements to produce heavier ones. If you add up the masses of the individual particles that are fused, they are slightly greater than the final element that is produced – along the way, mass is converted into energy. (It's a little more complicated as the mass of atoms is itself mostly derived from energy, but however you look at it, $E=mc^2$ is involved.)

Experimental fusion reactors have been constructed for over 50 years, but we are only now within a few decades of them being potentially viable for energy production. Fusion reactors are green, giving off no greenhouse gases in the energy generation, but don't produce the high-level nuclear waste of a traditional fission reactor, and could be essential for future energy needs to top up and balance out renewable sources in a world that no longer uses fossil fuels.

Somewhat more fancifully, as the special theory tells us that motion slows time down, we could eventually have fast enough spaceships that they could make significant journeys into the future. The only problem is that they would be unable to return, as backward time travel is far more difficult.

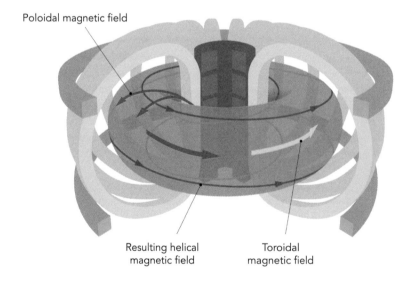

Poloidal magnetic field

Resulting helical
magnetic field

Toroidal
magnetic field

Above: The tokamak is a common design
for nuclear fusion reactors.

THE **ESSENTIAL** SUMMARY

ORIGINS	KEY THEORIES AND EVIDENCE	CRITICS	WHY IT MATTERS	FUTURE DEVELOPMENTS
1638 Galileo writes *Discorsi e dimostrazioni matematiche intorno a due nuove scienze*, including his concept of relativity. He points out that when we say something is moving, we have to ask: with respect to what?	According to Galilean relativity, all **motion is relative**. There is no absolute frame of reference.	Though Newton mostly worked with relative space and time he believed there were **absolute values** and used the example of a **rotating bucket of water** to try to prove this.	Galilean relativity is important in understanding **relative motion**: for example, in car crashes and planes flying through moving air.	**Fusion reactors**, which generate energy by fusing atomic nuclei together and producing energy from lost mass are very important for the future of green energy production.
1687 Isaac Newton's masterpiece *Philosophiæ Naturalis Principia Mathematica* is published, in which, despite dealing mostly with relative space and time, he argues for absolute space and time as well.	Einstein added in the fact that the **speed of light has to be constant** (in the same medium), however you move with respect to it. The result of Einstein's work was to show that time and space are linked and not absolute. When an object is moving its **time slows**, its **mass increases** and it **shrinks in the direction of movement**.	The special theory of relativity was **not widely criticized** – but it was also not widely recognized until the **general theory** was added.	The special theory of relativity becomes important when objects are moving at **extremely high speeds**: for example, in **particle accelerators**. The special theory is also the **jumping-off point** for the **general theory**, which would explain gravity.	With fast enough spaceships, the special theory makes **time travel** into the future possible.
1905 Albert Einstein builds on work by Lorentz and FitzGerald to present his special theory of relativity.	An implication of the special theory is that events at separate locations that appear simultaneous **will not be simultaneous** to a moving observer.		The implications of $E=mc^2$ are highly significant both for theoretical and practical uses in nuclear physics.	
1905 Later the same year, Einstein publishes another paper, deriving from the special theory that $E=mc^2$.	A simple extension of the special theory shows that **mass and energy are equivalent**.			

THE COMPONENTS OF EVERYTHING: FORCES AND PARTICLES

THE **ESSENTIAL** IDEA

"IT WILL BE FOUND THAT EVERYTHING DEPENDS ON THE COMPOSITION OF THE FORCES
WHICH THE PARTICLES OF MATTER ACT UPON ONE ANOTHER; AND FROM THESE
FORCES, AS A MATTER OF FACT, ALL PHENOMENA OF NATURE TAKE THEIR ORIGIN."

ROGER BOSCOVICH, 1758

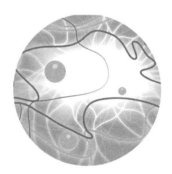

THE THREE FORCES

The essential components of reality are forces and particles. Particles divide into two in the "standard model" of particle physics: fermions and bosons. Fermions are the matter particles – in effect, "stuff". Meanwhile, bosons are responsible for interactions between matter particles – they are called "force carriers" because they move between fermions, carrying the forces of nature.

We have already met the electromagnetic force, carried by photons. Two other forces act on the nuclei of atoms. The strong force holds the component particles of atomic nuclei together, while the weak force is involved in nuclear reactions. The strong force is carried by bosons called gluons, while the weak force requires three bosons: Z bosons and positively and negatively charged W bosons.

The final force is a loner which we will look at separately – gravity. If gravity were brought under the umbrella of quantum physics, its boson would be the graviton, but as yet we don't have a theory to do this. And there's also a final particle, the Higgs boson, a disturbance in the field that gives other particles mass. Two other concepts, dark matter and dark energy, are covered in a separate section.

ORIGINS

The idea of matter being made of particles goes back to the ancient Greek philosophers Leucippus and Democritus from the later part of the fifth century BC. They argued that matter was made of indivisible particles. This was a logical conclusion of the idea of cutting something into smaller and smaller pieces – eventually a piece would be so small that it would be impossible to cut further. The Greek for "uncuttable", *atomos*, gives us the name atom. However, this early atomic theory was disputed by Aristotle, who won the argument, relegating atomism to a minority concept for over 2,000 years.

Newton would argue that light was made of corpuscles, and, with a number of other philosophers, he felt that atoms were real, but this also went against the mainstream of his time. It wasn't until the start of the nineteenth century that English scientist John Dalton used the idea of atoms as real objects to explain the structure of chemical compounds, based on the relative weights of elements. Atoms gradually become more accepted, as they were necessary for the statistical mechanics approach to gas theory, and became fully accepted by the start of the twentieth century.

"NEWTON WOULD ARGUE THAT LIGHT WAS MADE OF PARTICLES HE CALLED CORPUSCLES, AND, WITH A NUMBER OF OTHER PHILOSOPHERS, HE FELT THAT ATOMS WERE REAL, BUT THIS ALSO WENT AGAINST THE MAINSTREAM OF HIS TIME."

The early 1900s was also the period when atoms were shown to have structure, containing other, smaller particles. First, the electron was discovered, emitted by atoms, by English physicist J. J. Thomson. Soon after, in experiments led by the New Zealand physicist Ernest Rutherford, the atom was found to have an internal structure with a small, heavy central part, given the name "nucleus" after the name of the central part of a biological cell.

The current model of particle physics was gradually developed through the twentieth century, and the forces of electromagnetism and gravity were joined by the weak and strong nuclear forces. The final piece of the jigsaw (for the moment), the Higgs boson, was theorized in the 1960s and its existence was confirmed experimentally in 2012.

Opposite: Boson particles carry the forces than bind matter particles together.
Top: Ernest Rutherford (1871–1937).
Bottom: John Dalton (1766–1844).

KEY **THEORIES** AND **EVIDENCE**

THE STANDARD MODEL AND FOUR FORCES

"MOST OF THE FUNDAMENTAL IDEAS OF SCIENCE ARE ESSENTIALLY SIMPLE, AND MAY, AS A RULE, BE EXPRESSED IN A LANGUAGE COMPREHENSIBLE TO EVERYONE."

ALBERT EINSTEIN, 1938

The standard model of particle physics comprises 17 particles. Twelve of these are matter particles, known as fermions, though the vast bulk of matter that we experience is made up of just three types of particle: up and down quarks plus electrons. A combination of two up quarks and one down makes a proton, while two downs plus an up makes a neutron – the more familiar compound particles of the atomic nucleus.

QUARKS

Quarks can also combine in pairs consisting of a quark and an antiquark to form short-lived matter particles called mesons. The four other types of quark – charm, strange, top and bottom – take part in high-energy particle reactions but are rare in nature. Quarks are joined in the fermion category by the six leptons – these include the familiar electron, but also muons and tau particles, which are effectively extra-massive versions of electrons, plus three varieties of neutrino, a very light particle with no electrical charge, which is emitted during nuclear reactions.

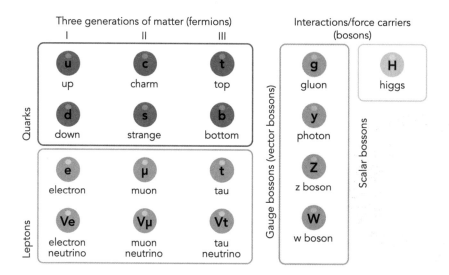

Above: The standard model shows the roles of the 17 fundamental particles.
Opposite top: The combination of quarks in a proton and a neutron.
Opposite bottom: A particle and its antiparticle meeting annihilate to pure energy.

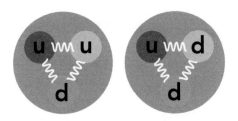

We then have the bosons, responsible for three of the fundamental forces. The most familiar is the photon, the force carrier for electromagnetism, which also makes up light. Then there is the gluon, which ties quarks together to form protons, neutrons and mesons. The Z boson and W bosons, carrying the weak force, are unusual in being force carrier bosons with mass, and in the case of the W can be positively or negatively charged. Bringing up the rear is the Higgs boson, a disturbance in the Higgs field, the field filling all of space that gives particles their mass.

"EACH MATTER PARTICLE HAS AN EQUIVALENT ANTIPARTICLE, A MIRROR IMAGE EQUIVALENT WITH SOME REVERSED PROPERTIES, INCLUDING ELECTRICAL CHARGE."

Each matter particle has an equivalent antiparticle, a mirror image equivalent with some reversed properties, including electrical charge. So, for example, an antiquark has the opposite charge to a quark, and an antielectron (also known as a positron) has the opposite charge to an electron. Neutral matter particles still have antiparticles, but with more obscure properties reversed. Neutral bosons are often described as being their own antiparticle. Particle/antiparticle pairs can be created from energy, and when a particle meets its antiparticle the two annihilate, turning back to pure energy in the form of photons.

The section on electromagnetism covers the electromagnetic force, and the section on gravity the gravitational force. The strong nuclear force both holds quarks together to form particles such as protons and neutrons and effectively leaks out of these to hold the nucleus of an atom together, even though the positively charged protons repel each other. The weak nuclear force is responsible for some nuclear decay processes and the ability of a neutrino to switch types.

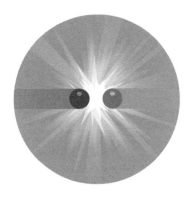

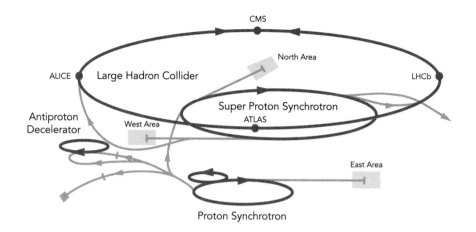

CMS

North Area

ALICE • Large Hadron Collider

LHCb

Super Proton Synchrotron
ATLAS

Antiproton
Decelerator

West Area

East Area

Proton Synchrotron

EVIDENCE

The picture of the standard model was built in part from a theory to explain the zoo of particles that were discovered through the twentieth century, but then confirmed by measurements made with particle accelerators which collide particles at near light speed to create a whole host of new, short-lived particles.

The electron was the first member of the standard model isolated, while the existence of some of the others was deduced before their particles were discovered. For instance, the existence of neutrinos was theorized due to mass that seemed to be disappearing during nuclear decay.

Antimatter is produced when there is sufficient energy in a collision to produce a matched pair of matter and antimatter particles – the first to be detected was the positron, behaving like an electron, but its opposite charge caused it to curve in the opposite direction as it moved through an electromagnetic field.

"THE ELECTRON WAS THE FIRST MEMBER OF THE STANDARD MODEL ISOLATED, WHILE THE EXISTENCE OF SOME OF THE OTHERS WAS DEDUCED BEFORE THEIR PARTICLES WERE DISCOVERED."

Above: The Large Hadron Collider at CERN is a complex structure centred on a 17-mile tunnel.
Opposite: The CMS experiment in the Large Hadron Collider is nearly 50 feet across.

CRITICS

The earliest critics of atoms were ancient Greek philosophers led by Aristotle, who believed the existence of atoms was impossible as it would imply a totally empty void between them, which was considered philosophically impossible. Even in the nineteenth century, some leading physicists such as William Thomson, Lord Kelvin thought that atoms were just a useful concept for calculations, but didn't actually exist. Clearly there was something there, but it did not have to be a particle in the true sense. Kelvin believed that atoms were really vortexes in the ether.

The existing standard model of particle physics works well, but has a few gaps. Some believe that it is missing half its particles, because they think that each particle has a "supersymmetric" partner – each fermion with a boson partner and each boson with a fermion. The bosonic partners of the fermions are named by prefixing with an "s-" – so quark is partnered by a squark and an electron by a selectron – while the fermionic partners of bosons get an "-ino" ending, giving a photon a partner called a photino and so on. Such supersymmetric particles are required for some versions of string theory, the most widely supported attempt to combine quantum theory and gravitation, but as yet none has been detected, even though some should be theoretically discoverable by the Large Hadron Collider. It seems increasingly likely that supersymmetric particles do not exist.

WHY IT **MATTERS**

The standard model is our most comprehensive approach to describe the nature of matter and how the forces that control the interaction of matter particles work. The model helps us understand both the basics of how matter is put together and many aspects of astronomy and cosmology, from the processes that take place within stars to the earliest development of the universe.

Unlike many other aspects of physics, there is little in the way of practical application for the understanding of particles – although these are quantum particles, and, as we have seen, quantum physics is crucially important for modern economies and technology. Having said that, work with particle accelerators, which use electromagnetic fields to accelerate charged particles, has had spin-off benefits for medical science. Of the forces, electromagnetism (given its own section) has by far the biggest practical significance to everyday life.

Being able to fill in the gaps in the standard model – whether through supersymmetry or some other restructuring – is likely to be an essential step towards a big picture of how quantum physics and the general theory of relativity (explaining gravity), which are currently incompatible, come together. The standard model works well, but we know that there has to be more to come.

FUTURE **DEVELOPMENTS**

At the moment we are at an impasse on the future of the standard model. Our existing knowledge of the model is very accurate, but there are gaps – for example, the model predicts that neutrino particles should not have mass, which was assumed to be the case for some decades. Some physicists still expect that if we can build even more powerful particle accelerators (at the time of writing, CERN is attempting to justify a multi-billion-dollar replacement of the Large Hadron Collider) we will be able to detect supersymmetric particles, transforming the standard model, but others believe this is a waste of money, as such particles should already have been detected.

"THE GENERAL THEORY OF RELATIVITY IS INCOMPATIBLE WITH QUANTUM THEORY."

On the forces front, the biggest failing we have is that three of the fundamental forces, electromagnetism and the strong and weak nuclear forces, are quantized (they are consistent with quantum physics). However, the fourth, gravity, is not. The general theory of relativity is incompatible with quantum theory. Attempts to provide a quantum theory of gravity, including string theory and loop quantum gravity, have as yet failed to provide any evidence for their correctness. The search goes on and it is quite possible that quantum gravity will at some time in the future provide a better model of reality.

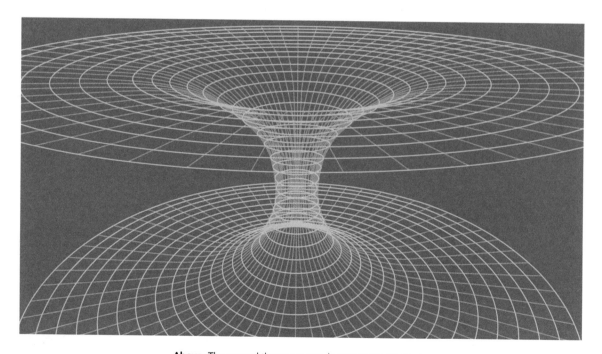

Above: The general theory accurately represents gravity as warped spacetime, but is incompatible with quantum theory.

THE **ESSENTIAL** SUMMARY

ORIGINS	KEY THEORIES AND EVIDENCE	CRITICS	WHY IT MATTERS	FUTURE DEVELOPMENTS
5th century BC Ancient Greek philosopher Leucippus and his pupil Democritus suggest that matter is made up of tiny particles: atoms.	We understand the behaviour of matter through four **fundamental forces** and 17 **particles**.	Aristotle believed **atoms could not exist** as there would have to be a void between them, which he thought was impossible.	The standard model gives us our **best picture** of how matter is constructed and how forces enable matter particles to interact.	At the moment there is an **impasse** on the future of the standard model.
1808 John Dalton publishes a first detailed description of a modern atomic theory of the elements.	The **standard model** of particle physics consists of **fermions** (matter particles) and **bosons** (force carriers).	Some physicists **did not accept the existence of atoms** until the start of the twentieth century.	Though there are limited practical applications, **particle accelerators** have medical technology spin-offs and **electromagnetism** is hugely important to technology.	We can expect to see a **refinement of the standard model**.
1897 J. J. Thomson demonstrates the existence of the electron.	All matter particles have an **antiparticle** with the reverse values of some crucial properties.	Some physicists believe each particle in the standard model should have a **supersymmetric partner** – but they have never been detected.		Filling in the gaps in the standard model will be an essential step to gaining a **big picture** of how quantum theory and gravity fit together.
1911 Ernest Rutherford demonstrates atomic structure.	The **four forces** are gravity, electromagnetism and the strong and weak nuclear forces.			

GRAVITY
(MORE TIME AND SPACE)

THE **ESSENTIAL** IDEA

"WHILST HE WAS MUSING IN A GARDEN IT CAME INTO HIS THOUGHT THAT THE POWER OF GRAVITY (WHICH BROUGHT AN APPLE FROM THE TREE TO THE GROUND) WAS NOT LIMITED TO A CERTAIN DISTANCE FROM THE EARTH BUT THAT THIS POWER MUST EXTEND MUCH FARTHER THAN WAS USUALLY THOUGHT. WHY NOT AS HIGH AS THE MOON SAID HE TO HIMSELF..."

JOHN CONDUITT (REPORTING A CONVERSATION WITH ISAAC NEWTON), 1726

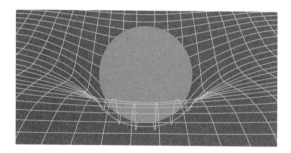

One of the four fundamental forces of nature, gravity, is extremely weak compared to the other three, only having a noticeable effect when a relatively massive body is involved. Gravity keeps things in place on the Earth – because the Earth is a very large body – and gravity keeps moons in orbit around planets, planets around stars and stars in galaxies. Although it drops off very quickly, strictly speaking gravity has no limit.

Gravity is an effect of objects with mass, which distort the spacetime around them. As a result of this warping of space, moving objects that naturally travel in a straight line pass along a curved trajectory, taking them around the massive body in an orbit. The warping of time causes a stationary object near to a massive body to accelerate towards it.

The solutions to the equations of the gravitational theory, the general theory of relativity, suggest the existence of black holes and allow for gravitational waves to cross the universe.

PHYSICS AND COSMOLOGY - GRAVITY (MORE TIME AND SPACE)

ORIGINS

The ancient Greek approach to gravity, formulated by Aristotle in the fourth century BC, was to pair it with levity. Gravity was the natural tendency of earth and water to head for the centre of the universe, while levity was the natural tendency of air and fire to move away from it.

Our modern idea of gravity owes much to Isaac Newton. While an apple didn't fall on his head, in his old age he did recall seeing an apple fall at his family home in England, inspiring him to think that the same force that caused an apple to fall also kept the Moon in its orbit. In a feat of mathematics in his book *Philosophiæ Naturalis Principia Mathematica*, Newton described the workings of gravity as an attraction between bodies with mass, dependent only on the two bodies' masses and the distance between them. Newton put forward no hypothesis for how gravity could work (though he believed it was due to some kind of flow of particles).

"OUR MODERN IDEA OF GRAVITY OWES MUCH TO ISAAC NEWTON. WHILE AN APPLE DIDN'T FALL ON HIS HEAD, IN HIS OLD AGE HE DID RECALL SEEING AN APPLE FALL AT HIS FAMILY HOME IN ENGLAND"

While Newton's mathematics still holds today, except in extreme circumstances, his work was effectively completed by Albert Einstein in 1916 with the general theory of relativity. This provided an explanation for how gravity acted at a distance, as it relied on the idea that matter warped space and time, and it was this warping that influenced how bodies would move through spacetime.

The general theory is mathematically complex. Early in its development, English physicist Arthur Eddington was asked if it were true that only three people in the world understood Einstein's theory. Eddington is said to have replied: "Who is the third?" But the principles were relatively straightforward – it was only the mathematics that proved a struggle for many physicists. The general theory made several predictions that were different from the expected results of Newton's theory. It was Eddington who led an expedition to observe a solar eclipse in 1919 that would confirm Einstein's theory and help make Einstein the most famous scientist of all time.

Opposite: Massive bodies, such as the Earth, warp space and time, causing straight line paths to curve.
Top: Isaac Newton (1643–1727).
Bottom: Arthur Eddington (1882–1944).

KEY **THEORIES** AND **EVIDENCE**

INVERSE SQUARE LAW AND WARPING SPACETIME

"BUT THE MOST IMPRESSIVE FACT IS THAT GRAVITY IS SIMPLE. IT IS SIMPLE TO STATE THE PRINCIPLES COMPLETELY AND NOT HAVE LEFT ANY VAGUENESS FOR ANYBODY TO CHALLENGE THE IDEAS OF THE LAW."

RICHARD FEYNMAN, 1967

$$F = G\, \frac{m_1 m_2}{r^2}$$

Although the equation doesn't appear in his writing, Newton's gravitational theory boils down to there being a gravitational force of attraction between two bodies of masses m_1 and m_2 of Gm_1m_2/r^2 where G is a constant value and r is the distance between the two bodies. From this, it can be seen that the attraction goes up as the masses increase and as the square of the distance between them decreases. Another important implication that Newton brought out, bearing in mind that most objects aren't concentrated into a point, is that the strength acts as if all a body's mass is concentrated at its centre.

Although attempts were made, starting with Newton, to explain gravitational attraction as a result of pushes from an onslaught of invisible particles flying through the universe which were shielded by massive bodies, causing pressure, it would not be until Einstein that a theory that effectively explained gravity's ability to attract at a distance would be formulated. Building on his special theory of relativity, which excluded acceleration (hence the term "special", here meaning limited), Einstein started by showing that gravity and acceleration were indistinguishable.

A simple example of this would be what happens inside a spaceship without windows. If it sits on a body like the Earth with a strong gravitational pull, those

inside feel a force towards the bottom of the ship. But if it steadily accelerates through space, they feel an identical acceleration. If a beam of light is shone across the inside of the ship during acceleration, it will bend – so, equally, a beam of light crossing the ship in a gravitational field should bend. The mathematics proved complex, but Einstein was able to put together a collection of equations, known as the field equations of gravity, which show how matter warps spacetime and how warped spacetime causes the effects of gravitation.

The general theory would be used to predict many things about the universe, including its potential to expand and collapse, and the potential for stars to collapse into dimensionless "black holes" (a term introduced in the 1960s, though the prediction was made by German physicist Karl Schwarzschild while serving in the First World War in 1915, before Einstein had even completed his work).

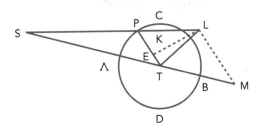

EVIDENCE

Newton supported his theory with page after page of geometry, proving his assertions. His original calculations had been made with the more sophisticated mathematics of calculus, which he had invented for the purpose (calling it "the method of fluxions"), but he mostly translated the results into geometry to make the outcome more acceptable to his readers. Newton's work made numerous predictions about orbits that matched observation, including being able to derive Kepler's laws of planetary motion, while the English scientist Edmond Halley, a major supporter of Newton who paid for the publication of the *Principia*, used Newton's theory to successfully predict the return of a comet, now named after Halley.

"BEFORE EINSTEIN WORKED OUT HIS EQUATIONS IT WAS KNOWN THAT NEWTON'S THEORY DIDN'T CORRECTLY PREDICT THE ORBIT OF MERCURY"

Before Einstein worked out his equations it was known that Newton's theory didn't correctly predict the orbit of Mercury, close enough to the Sun to be in a strong gravitational field. Einstein's equations matched observation perfectly. The theory also predicted that light from distant stars passing a massive body like the Sun would take a curved path, as the Sun warped spacetime around it. Usually stars that are visually close to the Sun are impossible to see, but during an eclipse they become visible – and their shift is as Einstein predicted. Since the first observations to show this in 1919, the general theory has proved to be accurate many times.

Opposite top: The equation for the force of gravity according to Newton's law.
Opposite bottom: Einstein realized that acceleration and gravity were indistinguishable.
Above: Newton translated much of the working in his masterpiece from calculus to geometry.

CRITICS

Although Newton's mathematics produced very little argument, there was considerable doubt about Newton's use of the term "attraction" to represent the force between the Earth and the Moon, for example. We would now see this as a perfectly normal use of the word, but at the time it would only have applied in the sense of finding another person attractive, suggesting some form of romantic attraction between the Earth and the Moon. Newton was criticized for making gravity an "occult" (hidden) force that somehow reached out and influenced things at a distance, where all known phenomena acting remotely required something to pass from A to B to cause the influence.

"NEWTON WAS CRITICIZED FOR MAKING GRAVITY AN "OCCULT" (HIDDEN) FORCE"

Many struggled with the mathematics of the general theory of relativity, including Einstein. This was because it moved away from the geometry of flat space to curved space, a relatively recent mathematical development. Einstein's early work on the subject contained an error, and one of the greatest mathematicians of the era, the German David Hilbert, raced Einstein to complete a correct series of equations. Luckily for Einstein, he pipped Hilbert to the post. There have been many attempts to find flaws with Einstein's thinking – he has proved a magnet for pseudo-scientists and amateurs who are always "disproving Einstein" – but no serious scientist has found a flaw in his work on the general theory.

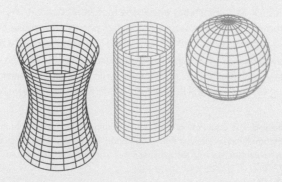

WHY IT **MATTERS**

It is self-evident that gravity matters, in both its obvious roles of keeping things in place on the Earth and in maintaining astronomical orbits. Evidence from those living in microgravity on the International Space Station also points to the necessity of gravity for many living things. Plants struggle to grow as they rely on gravity for the direction of root development, while eggs fail to hatch, and humans suffer bone deterioration and problems with organs floating up inside the body.

Understanding gravity at the Newtonian level is also important for anything involving movement through a gravitational field, whether for ballistic considerations and flights in the atmosphere or for space travel, where, for example, an understanding of the interaction of probes with gravitational fields is often used in a so-called "slingshot" effect, where a probe flies around a planet or the Sun and picks up speed in the process.

At the level of the general theory of relativity, the benefits are more in terms of knowledge – getting a better understanding, for example, of the way that universe has developed and will do so into the future, and understanding unusual structures such as neutron stars and black holes.

Opposite top: David Hilbert (1862–1943).
Opposite bottom: Space can be flat or curved inwards or outwards.
Above: The International Space Station has been used to experiment on the effects of absence of gravity on life.

FUTURE **DEVELOPMENTS**

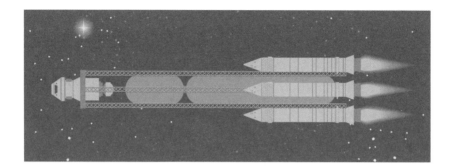

As we have already seen in forces and particles, one of the biggest problems physics faces is that gravity is not quantized: the general theory of relativity depends on continuously variable values, which are incompatible with quantum theory. Attempts to provide a quantum theory of gravity, including string theory and loop quantum gravity, have as yet failed to provide any evidence for their veracity. The search will go on and it is quite possible that quantum gravity will at some time in the future be demonstrated to be the reality.

If we are to spend more time in space – for example, travelling to Mars and further – it is likely that more work will be put into means of artificially generating the effects of gravity. There is no scientific basis for the kind of gravity generator that is common in science fiction, which like antigravity is highly unlikely to be possible. Instead, artificial gravity is likely to be generated by a mechanism using the equivalence of gravity and acceleration, either by having a ship that spends most of its journey under acceleration or by rotating a large enough segment of the ship that the inhabitants don't get dizzy.

Top: A spaceship keeping a constant acceleration would feel to have gravity.
Bottom: New approaches may enable us to produce a quantum theory of gravity.

THE **ESSENTIAL** SUMMARY

ORIGINS	KEY THEORIES AND EVIDENCE	CRITICS	WHY IT MATTERS	FUTURE DEVELOPMENTS
4th century BC Aristotle pairs gravity with levity. Gravity was a tendency of some elements to head for the centre of the universe and levity a tendency to move away from it.	Newton's gravitational theory makes the gravitational force between bodies of masses m_1 and m_2 to be Gm_1m_2/r^2 where G is a constant value and r is the distance between the two bodies.	Newton received some criticism for his use of the term "**attraction**" and making gravity an "occult" or hidden force.	Self-evidently gravity matters in **keeping things in place** on Earth and keeping bodies in orbit.	Theoretical physics hopes to find a workable **quantum theory of gravity** so that it can be integrated with the other fundamental forces and simplify our understanding of the universe.
c1666 Newton is said to have observed the apple falling that caused him to think that the same force causing the apple to the fall also keeps the Moon in orbit.	Einstein observed that acceleration and the force of gravity are **equivalent** – they have exactly the same effect.	Einstein was **nearly beaten** to a solution by mathematician David Hilbert, after Einstein's first effort contained an error.	**Living things struggle** if they spend too much time without gravity.	If we are to spend longer in space, we need **workable artificial gravity**, either from constant acceleration or from rotating part of a ship (also causing acceleration).
1916 Albert Einstein completes his general theory of relativity, both refining Newton's predictions and providing a cause for the force of gravity as matter warps spacetime.	Einstein puts together the field equations of gravity, which show how **matter warps spacetime** and how warped spacetime causes the effects of gravity.		An understanding of Newtonian gravity is essential for **ballistics, flight and space travel**.	
1919 Eddington leads an expedition to observe the solar eclipse, confirming Einstein's theory.	The general theory was used to predict the existence of **black holes** and the future of the universe.		The general theory of relativity is less practical, but essential for **understanding aspects of the universe**.	

THE BIG BANG

THE **ESSENTIAL** IDEA

"THE UNIVERSE CAME INTO BEING IN A BIG BANG, BEFORE WHICH, EINSTEIN'S THEORY INSTRUCTS US, THERE WAS NO BEFORE. NOT ONLY PARTICLES AND FIELDS OF FORCE HAD TO COME INTO BEING AT THE BIG BANG, BUT THE LAWS OF PHYSICS THEMSELVES..."

JOHN WHEELER, 1982

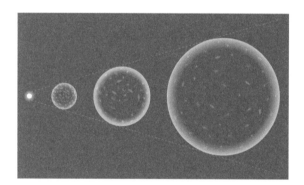
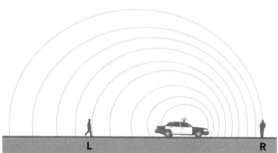

The idea of the Big Bang originates from the discovery that the universe is expanding. The matter of the universe is not expanding into previously empty space: space itself is growing. The discovery was made as a result of observing the redshift of galaxies – the optical equivalent of the Doppler effect. This showed that, with a few local exceptions, every galaxy in the universe is moving away from us. And this effect would arise if the universe itself were expanding.

If we imagine a movie of the expanding universe and run it backwards – the equivalent of looking backwards in time – we would see a smaller and smaller universe. While expansion into the future could, in principle, carry on forever, when we look back in time to a shrinking universe there has to come to a point where it

Above left: In the expanding universe, space itself expands.
Above right: The Doppler effect means that waves are closer together and further apart behind a moving source.
Opposite top: Henrietta Swan Leavitt (1868–1921).
Opposite bottom: Edwin Hubble (1889–1953).

"EVERY GALAXY IN THE UNIVERSE IS MOVING AWAY FROM US"

could be no smaller, where the universe has shrunk to nothing. This point in time, around 13.8 billion years ago, can be considered the origin of the universe. In the Big Bang theory – the best accepted theory of the origin of the universe – there is no "before" – this is the beginning of space and time in our universe.

ORIGINS

There had been many mythological explanations for the origins of the universe, usually involving creation by a deity at some point in the past or an infinite existence, but there was no scientific explanation available until the expansion of the universe was observed.

DYNAMIC UNIVERSE

When Einstein produced his general theory of relativity in 1916, simplified solutions of his equations were produced for the universe as a whole which suggested that the universe should either expand or contract. Einstein thought this did not match reality and added a factor known as the cosmological constant, which allowed for a stable universe. He later referred to this as his greatest mistake and removed it.

In 1922, the Russian physicist Alexander Friedmann completed the solution to the equations for an expanding universe. Later in the decade, the American astronomer Edwin Hubble had provided evidence that the universe was far bigger than just the Milky Way with the first measurements of distances to other galaxies. The measurement used was based on variable stars, which go through a repeated cycle of growing brighter and dimmer.

The discovery these could be used had been made in 1912 by American astronomer Henrietta Swan Leavitt, who published a paper showing that the maximum brightness of variable stars had a simple relation to the period of time it took them to pass through a cycle of intensity. Knowing their actual brightness, apparent differences in their intensity had to be due to their distance from us. It meant that the stars could be used as "standard candles" to measure far greater distances than was possible with the basic method of distance measurement, known as parallax, based on the apparent movement of the star in the sky when the Earth travels round its orbit.

With the exception of our nearest neighbour, the Andromeda galaxy, which had a blueshift, galaxies were mostly redshifted, and Belgian scientist Georges Lemaître proposed what would become known as Hubble's law, that the size of the redshift increased with distance. The inference which Lemaître drew was that the universe was expanding and had been smaller in the past, starting as what he would refer to as a primeval atom or "cosmic egg".

KEY **THEORIES** AND **EVIDENCE**

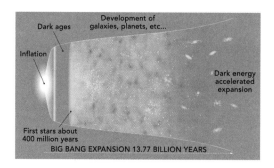

Dark ages

Inflation

Development of
galaxies, planets, etc...

Dark energy
accelerated
expansion

First stars about
400 million years

BIG BANG EXPANSION 13.77 BILLION YEARS

BEGINNINGS, EXPANSION AND INFLATION

According to basic Big Bang theory, in the very beginning, space and time (along with the laws of physics) came into being (with no known cause). An inconceivably small time after this came the Big Bang, when the universe began to expand. As it continued to get bigger, tiny variations imposed by the uncertainty principle resulted in larger variations that would eventually be detected in the cosmic microwave background radiation and produced the variations in density that would form galaxies.

TIMELINE OF THE UNIVERSE

After around 10^{-36} seconds, for no obvious reason, the theory assumes that the universe underwent inflation: a sudden, drastic acceleration in the rate of expansion that enabled the universe to increase its volume by at least a factor of 10^{78} before settling back to a more conventional rate of expansion at around the 10^{-32} second mark.

By this point, the universe had gone from pure energy to some matter particles existing in the form of quarks, antiquarks and the gluons that would eventually link them. Because energy produces pairs of particles – a particle and its antiparticle, there should have been as much antimatter as matter. Where all the antimatter went is one of the significant mysteries of cosmology. By 10^{-12} seconds in, leptons such as electrons (and their antiparticles) had formed, and between 10^{-6} and 1 second of the life of the universe, protons and neutrons were forming from the quarks.

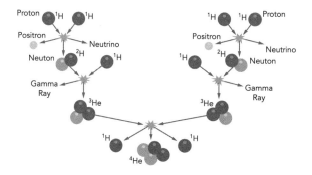

Initially, the only chemical element present would be hydrogen (a proton is a hydrogen nucleus), but there was a period when the whole universe acted like a star, fusing hydrogen into helium and a small amount of lithium. This process would have happened between about 2 and 20 minutes in, by which time temperatures had dropped sufficiently for fusion to cease.

For the next 377,000 years, the cooling, expanding universe was an electrically charged plasma, preventing light from passing through it. But by the end of this period, temperatures had dropped sufficiently for atoms to form and the universe became transparent. As the universe continued to expand, gravity would gradually enable stars and galaxies to form a few hundred million years into the lifetime of the universe. Reality as we know it was on its way.

It ought to be stressed that although the Big Bang theory is the best-accepted theory which fits the evidence reasonably well (after being patched several times), it should not be taken as definitive fact, but rather the best guess given the current data.

EVIDENCE

As we have seen, it was the redshift of galaxies that provided the first evidence for the expansion of the universe. The most significant piece of evidence that would enable the Big Bang theory to relegate its main competitor to history was the cosmic microwave background. This source of low-level microwave radiation was discovered in 1964, arriving at pretty much the same intensity from every direction. It is the radiation that first crossed the universe when it was 377,000 years old and became transparent to light. Further supporting evidence came as it was discovered that very distant galaxies (hence a long time in the past, as light takes time to reach us) were quite different from current ones, implying that the universe had evolved rather than continued forever in the same way, as suggested by the Big Bang's main competitor, the steady state model.

Inflation is more a patch to fix a problem with the Big Bang theory than something supported by evidence. The universe seemed to be too uniform. It appears too big for distant regions to have been in contact after early quantum fluctuations produced differences in temperature if the basic Big Bang theory were true, yet they are physically so similar that they must have been co-located at some point in their history. As a result, the ultra-fast expansion of inflation was suggested.

Opposite top: The timeline of the expanding universe.
Opposite bottom: The nuclear fusion process from hydrogen to helium.
Above: The cosmic microwave background radiation image from the WMAP satellite.

CRITICS

"WE NOW COME TO THE QUESTION OF APPLYING THE OBSERVATIONAL TESTS TO EARLIER THEORIES. THESE THEORIES WERE BASED ON THE HYPOTHESIS THAT ALL THE MATTER IN THE UNIVERSE WAS CREATED IN ONE BIG BANG AT A PARTICULAR TIME IN THE REMOTE PAST."

FRED HOYLE, 1948

Ironically, it was one of the Big Bang theory's leading opponents who named it. English astrophysicist Fred Hoyle along with Austrian-British cosmologist Hermann Bondi and Austrian-British-American physicist Thomas Gold were uncomfortable in part with the religious overtones that the Big Bang theory was sometimes endowed with and produced a strongly competing theory in the steady state model, which proposed a permanent expansion and creation of matter on a small scale, with no beginning. Evidence would result in the dismissal of steady state theory (Hoyle argued this was more a sociological decision than one based on the data) without fixing it to match observation, as happened with Big Bang theory.

One such enhancement of Big Bang theory that remains strongly criticized among some physicists is inflation, for which there is no good evidence yet. In 2014, an experiment called BICEP2 announced what was thought to be evidence of inflation, but soon after its findings were interpreted as being due to the influence of interstellar dust.

WHY IT **MATTERS**

In a sense, the Big Bang is of little importance. Knowing how the universe began makes no difference to our everyday lives. It isn't going to result in any breakthrough technology, or practical application. And yet the very fact that every civilization has tried to explain how the universe came into being shows that we have a deep fundamental need to understand where the universe (and hence we) came from.

"IF BIG BANG THEORY IS CORRECT, THE VERY EARLY UNIVERSE WAS SO SMALL THAT QUANTUM EFFECTS HAD A MAJOR IMPACT ON ITS EXISTENCE"

The Big Bang is strongly tied in with our understanding of gravity and quantum physics. If Big Bang theory is correct, the very early universe was so small that quantum effects had a major impact on its existence – but as yet we have no way of integrating quantum physics and the general theory of relativity, so there is a strong uncertainty over exactly what happened in these earliest of times.

Because of the dependence on indirect measurements, plus the gaps in our understanding, cosmology remains the most speculative of the sciences, and arguably could be seen as the closest thing we still have to the ancient philosophers' viewpoints. Cosmology is, scientifically speaking, a fascinating living fossil.

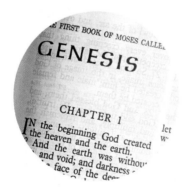

Opposite top: Fred Hoyle (1915–2001).
Opposite bottom: The BICEP2 observatory near the South Pole.
Above: Opening of the Bible – Genesis – an early explanation for the origin of the universe.

FUTURE **DEVELOPMENTS**

One of the biggest problems we face in dealing with cosmology is the inability to see back to when the universe was less than 377,000 years old, the period when light was incapable of crossing the opaque universe. Historically our methods of looking out into the universe have depended on different energies of electromagnetic radiation – everything from radio to gamma rays as well as the more familiar visible light. But we now have two options, gravitational waves and neutrinos, which could be observed from the earlier universe. In each case, detection is in its infancy, but in the future these methods could give us a better picture of the beginning.

It would also help if we could fill in the gaps in the standard model of particle physics, get a better understanding of dark matter and dark energy (more on these soon) and find a way to unify gravity and quantum physics to have the physical tools to be able to get a better handle on the early universe. It seems likely we will still see a variation of the Big Bang, but our current theory is almost certainly wrong in some aspects.

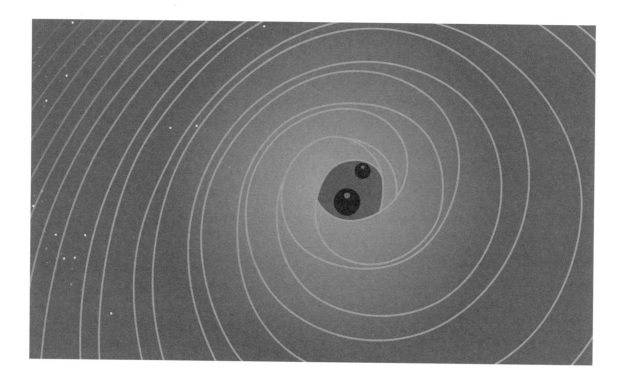

Above: Gravitational waves, generated by events such as merging black holes, could give insights into the early universe.

THE **ESSENTIAL** SUMMARY

ORIGINS	KEY THEORIES AND EVIDENCE	CRITICS	WHY IT MATTERS	FUTURE DEVELOPMENTS
Throughout recorded history attempts have been made to explain where the universe came from, usually involving creation by a deity or infinite existence. **1916** Einstein's general theory of relativity suggests the universe should expand or contract. **1922** Alexander Friedmann solves general relativity for an expanding universe. **Mid-1920s** Edwin Hubble shows other galaxies exist and are mostly redshifted. **1927** Georges Lemaître proposes a universe that has expanded from a primeval atom.	Space and time begin as an **infinitesimally small point** that begins to expand with the Big Bang. In a tiny fraction of a second, in the process called **inflation**, the volume of the universe increases by a factor of 10^{78} or more. As the universe expands and cools, energy produces **matter/antimatter pairs**. Somehow only matter primarily remains in the observable universe. After the universe is 377,000 years old, light starts to cross it, which would become the **cosmic microwave** background. After a few hundred million years, gravity begins to form structures such as **stars and galaxies**.	Fred Hoyle, Hermann Bondi and Thomas Gold proposed an alternative **steady state model**, unhappy with the big bang. This would be dismissed without the enhancement to match evidence that the Big Bang received. Some physicists are highly critical of **inflation**, which is not based on evidence, but rather an arbitrary fix to Big Bang theory to match observation.	The Big Bang is of **no practical importance**, but developing cosmological theories seems to answer a deep-seated human need. Big bang theory is strongly linked to attempts to **unify gravity and quantum theory**: without this it is hard to make it complete. Arguably, cosmology is a **living echo** of the attempts of early philosophers to understand the universe.	**Gravitational waves and neutrinos**, could give us a more informed view of the early universe – detection of both is still in its infancy. As we **fill in gaps** in the standard model of particle physics and develop knowledge of dark matter and dark energy, and an understanding of the combination of gravity and quantum physics, we may get an improved equivalent of the Big Bang theory.

STARS, PLANETS, SOLAR SYSTEMS AND GALAXIES

THE **ESSENTIAL** IDEA

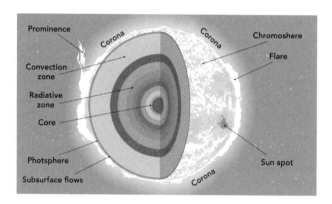

A million years or so after the Big Bang, the universe contained vast clouds of gas, primarily hydrogen. Over time the atoms would very slowly attract each other. Matter began to clump together, heating up from impacts with other atoms. In the most concentrated regions, the gas become denser, increasing temperature and pressure. Eventually this effect was strong enough to begin the process of nuclear fusion and the first stars were born.

These new bodies were spinning around, which pulled the material around them into a disk. Material in that disk also clumped together as the star originally did, forming gas planets, too small to have sufficient temperature and pressure to fire up as a star. After some of the first stars exploded as supernovas, interstellar space was filled not only with gas but dust of denser elements. When second-generation stars formed their solar systems, some planets would be rocky rather than gaseous.

Because the original matter in the universe was unevenly distributed, caused by early quantum fluctuations, stars emerged in large-scale clumps which developed their own identity through gravitational interaction: galaxies.

Above: Structure of the Sun.
Opposite top: Archimedes (287–212 BCE).
Opposite middle: Nicolaus Copernicus (1473–1543).
Opposite bottom: Our galaxy, the Milky Way, as seen from Earth.

ORIGINS

Ancient Greek astronomers divided the contents of the heavens into stars and "wandering stars" – planets – which initially were the Moon, Venus, Mercury, the Sun, Mars, Jupiter and Saturn. Most ancient cosmology thought that what we would now call the solar system was the entire universe, with the stars embedded in a sphere on the outside. In the third century BC, the Greek mathematician Archimedes calculated the size of the universe in an attempt to work out how many grains of sand it would take to fill it (this wasn't as silly as it sounds: it was an exercise in extending the number system, which at the time could not be practically used beyond 100 million). His estimate, around the diameter of the orbit of Saturn, was surprisingly good.

In the ancient model of the universe, each planet was fixed to a crystal sphere, enabling it to rotate around the Earth, as did the outermost sphere holding the fixed stars. Some thought this outer sphere was opaque with holes in it, and the stars were lit by a fire that surrounded the universe, shining through. The planets were thought to be similar to the Earth, to the extent that some speculated about living beings on the Sun.

THE COPERNICAN MODEL

When Nicolaus Copernicus put the Sun at the centre of the universe, rather than the Earth, a first step was made in making a wider universe possible. The main benefit of doing this was that it explained the strange paths of the outer planets, which appear to reverse in their tracks as a result of the interaction of the orbit of the Earth and the other planets.

With time it became clear that the stars were much further away than the planets and, with the power of telescopes to resolve more detail, it was realized that a band of light in the sky known as the Milky Way was a huge ribbon of stars. As a result, by the nineteenth century it was clear that the universe was far bigger than previously thought, making up what we now call our galaxy. It was only in the twentieth century that speculation by the eighteenth-century astronomer William Herschel that there were other galaxies beyond our own was confirmed and the main components of the universe recognized.

KEY **THEORIES** AND **EVIDENCE**

FUSION, PLANETARY DISKS AND GALACTIC DISTANCES

"I ALREADY HAVE THIS: THAT THE MOST TRUE PATH OF THE PLANET [MARS] IS AN ELLIPSE, WHICH DÜRER ALSO CALLS AN OVAL, OR CERTAINLY SO CLOSE TO AN ELLIPSE THAT THE DIFFERENCE IS INSENSIBLE."

JOHANNES KEPLER, 1605

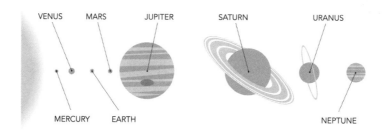

Our solar system consists of a star orbited by eight planets (Mercury, Venus, Earth, Mars, Jupiter, Saturn, Uranus and Neptune), along with a whole collection of minor planets and asteroids, comets and moons orbiting the various planets.

Like all stars, the Sun operates by nuclear fusion. Under the immense temperatures and pressures caused by the gravitational attraction of vast amounts of gas (the Sun contains about 99.9 per cent of solar system matter), the electrically charged atoms (ions) in the Sun are pushed sufficiently close together that, with a little help from quantum theory, the strong nuclear force is able to bind them to each other. They fuse into heavier ions, giving off energy – powering the rest of the solar system.

THE FORMATION OF THE SOLAR SYSTEM

When the solar system formed, clouds of gas and dust were attracted to the Sun. All this material was spinning: because the matter was not evenly distributed through space, there was more matter in some parts than others, causing an off-balance pull that started a rotation. As the material pulled inwards, angular momentum – the oomph of rotation – was conserved. As the material moved inwards, the rotation sped up, just as skaters' spin accelerates when they pull their arms in. Because of the spin, the matter in the solar system ended up in a disk rather than a ball – like a piece of pizza dough flattens out as it is spun around.

The nuclear fusion process makes heavier and heavier elements, starting by turning hydrogen into helium, until the process produces iron. A star can't make any heavier elements – to go further requires energy to be put into the system.

Above left: The planets, not to scale.
Above right: Conservation of angular momentum when a skater pulls her arms in.
Opposite top: Using parallax to work out the distance to a star.
Opposite bottom: The dark lines in a stellar spectrum, identifying different elements.

However, some types of star, nearing the ends of their lives, collapse inwards, causing an intense reaction that blows off their outer matter in an extremely bright event known as a supernova. This explosion provides the energy to generate heavier atoms, as does the collision of some kinds of star.

The size we believe the universe to be has grown with our understanding of its contents. Until the twentieth century it was thought that the solar system, and then the Milky Way, our local galaxy, was the whole universe. The Milky Way is around 200,000 light years across. (A light year is the distance light travels in 1 year, around 5.9 trillion miles (9.5 trillion kilometres)). However, we now know that the universe contains billions of galaxies, and is more than 90 billion light years across.

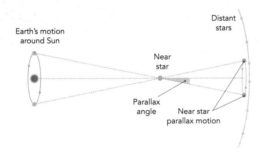

EVIDENCE

We know a surprising amount about the nature of different stars by the light they give off. When light shines through different elements they absorb particular energies of photons, leaving gaps in the spectrum. Spectroscopy – analysing spectra – means we can find out what elements are in a star and identify different types of star.

At first it was assumed that the Sun burned with a conventional fire, but this would only give it a lifetime of a few thousand years. It was only when nuclear reactions were understood that a mechanism allowing stars to last for billions of years was discovered.

The distance to closer stars can be measured using parallax – the same effect that makes close objects appear to "move" when viewed with first one eye and then the other. Knowing how far apart the eyes are, you can work out distances using the apparent movement. Astronomers view stars six months apart – from opposite sides of the Earth's orbit, producing a considerable parallax effect. Beyond the effective range of parallax measurement, "standard candles" are used. These are stars (and most recently supernovas) with a known brightness, making their relative brightness a measure of how far away they are.

We think the universe is 13.8 billion years old, so you might imagine we could only see objects 13.8 billion light years away, but because of the expansion of the universe the absolute limit is around 45 billion light years. However, the universe could be far bigger.

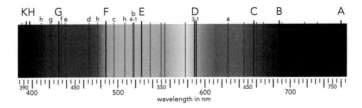

CRITICS

Famously, the idea that the Sun was the centre of the universe was strongly resisted in the sixteenth century. In part this was because Aristotle, whose physics required the Earth to be at the centre of things, was held in such high regard, and in part because the Bible mentioned the Sun moving. This provided the basis for Galileo's arrest (though he probably would not have been tried had he not insulted the pope in his book).

"EDWIN HUBBLE'S USE OF STANDARD CANDLES TO MEASURE DISTANCES TO OTHER GALAXIES WOULD PROVE THAT THEY WERE FAR BEYOND THE EXTENT OF THE MILKY WAY"

As late as 1920 there was considerable criticism of the idea that there was more to the universe than the Milky Way. On 26 April 1920, a debate was held at the Smithsonian in Washington, DC between astronomers Harlow Shapley and Heber Curtis. Shapley supported the view that nebulae – fuzzy patches of light in the night sky containing large numbers of stars – were clusters on the edge of the Milky Way, while Curtis believed they were galaxies in their own right. Astronomer Edwin Hubble's use of standard candles to measure distances to other galaxies would prove that they were far beyond the extent of the Milky Way. Hubble first put our nearest large neighbour, the Andromeda galaxy, about a million light years away, but we now know it is 2.5 million light years. (Hubble initially confused two different types of variable star used as standard candles.)

Above left: Harlow Shapley (1885–1972).
Above right: The trial of Galileo.
Opposite top: Barringer crater in Arizona.
Opposite bottom: The extent of the Chicxulub crater in the Gulf of Mexico.

WHY IT **MATTERS**

The primary value of understanding the main components of the universe is in having a better picture of where we live. Humans have a strong urge to explore and find out about our surroundings – and planets, stars and galaxies make up our universal environment.

Understanding the structure of the solar system is also valuable in terms of getting a better handle on asteroid impacts. Around 65 million years ago, a rocky object around 10 kilometres (6 miles) across crashed into the Earth at around 20 kilometres per second (12.5 miles per second). It produced a vast crater in the Gulf of Mexico, 200 kilometres (125 miles) across. This released a blast of energy five billion times as big as that from the nuclear bombs dropped during the Second World War. It wiped out all life around it and caused immediate, catastrophic climate change which resulted in the extinction of the dinosaurs.

Such impacts are rare – but will happen again at some time in the future. Having a better understanding of the structure and workings of the solar system enables us to be on the watch for other impactors and to develop plans to prevent them hitting the Earth.

Knowing about the basics of stars, planets and galaxies is also an essential starting point for getting a better understanding of the past and future of the universe as a whole.

FUTURE **DEVELOPMENTS**

Over time we have got better at making measurements to distant galaxies. It wasn't until the 1990s that we were able to use some types of supernova as standard candles to look out (and hence back in time) to galaxies where the light has been travelling towards us for over 10 billion years – over 70 per cent of the lifetime of the universe.

We still have more to discover about planetary formation. For example, it is now thought that the Earth's unusually large moon formed when a planet-sized body crashed into the young Earth, blasting out a large chunk that mixed material from the Earth and the collider. However, this is still not certain and there are some anomalies still to be resolved.

Inevitably, given the distances involved and the often-indirect measurements, as we get better space-based telescopes (which are far better than those on the Earth as they avoid the distortion caused by the atmosphere) we will be able to improve our knowledge. For example, it is only recently that we have been able to detect planets around other stars, but we are now on the verge of being able to find out about their atmospheres and composition. Similarly, new methods of seeing into space, such as using gravitational waves, are giving us the ability to detect what was never before possible.

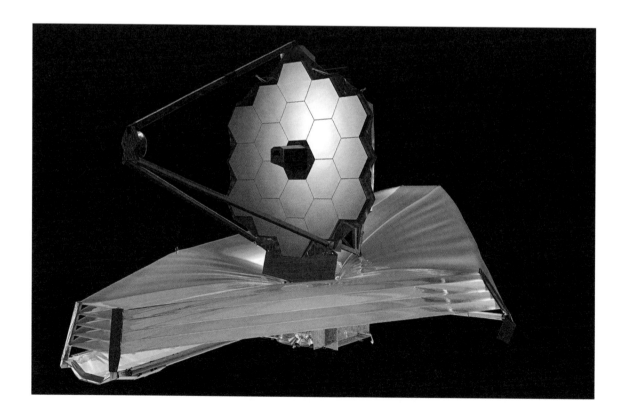

Above: The James Webb space telescope, launching in 2021.

THE **ESSENTIAL** SUMMARY

ORIGINS	KEY THEORIES AND EVIDENCE	CRITICS	WHY IT MATTERS	FUTURE DEVELOPMENTS
Ancient Greek astronomy divides the heavens into fixed stars and wandering stars (planets) which include the Sun and Moon but not the Earth. **3rd-century BC** mathematician Archimedes estimates the size of the universe as what we now know to be the diameter of the orbit of Saturn. **16th-century** astronomer Copernicus puts the Sun at the centre of the universe. **18th-century** astronomer William Herschel thinks that nebulae could be other galaxies.	The Sun (like all other stars) operates by **nuclear fusion**, releasing energy when atoms fuse together to make heavier elements. The **rotation of the material in the solar system**, started because it was unevenly distributed and accelerated as matter was pulled together, spinning the material out into a disk, like a pizza being spun. Nuclear fusion in stars made **elements up to iron**: heavier ones came from supernovas. The Milky Way is around **200 light years across**, but the universe contains **billions of galaxies** and is over 90 billion light years across. We know what stars are made of using **spectroscopy**, examining the spectrum of colours in the light they give off. Distances to stars are measured using **parallax** and **standard candles**.	Supporters of **Aristotle** argued the Earth was at the **centre of the universe** as his physics would not work without it. Biblical arguments were also used to support this position. As late as 1920 there was still disagreement over whether the **Milky Way was the whole universe**, or there were other galaxies. The dispute was won when distances to **other galaxies** were measured as being far beyond the size of the Milky Way.	Understanding our environment fits with our urge to **explore** and know our surroundings. Understanding the structure and mechanisms of the solar system enables us to **track and potentially deflect asteroids or comets** that could collide disastrously with the Earth.	Our ability to use **standard candles** to measure distances is improving all the time. We still have more to learn about **planetary formation** – for example aspects of the Earth/Moon system's development are still debated. Better space telescopes are enabling us to find out more about **planets around other stars**. New astronomical methods such as using **gravitational waves** are enabling us to see further (and hence further back in time).

DARK MATTER
AND DARK ENERGY

THE **ESSENTIAL** IDEA

"A COSMIC MYSTERY OF IMMENSE PROPORTIONS, ONCE SEEMINGLY ON THE VERGE OF SOLUTION, HAS DEEPENED AND LEFT ASTRONOMERS AND ASTROPHYSICISTS MORE BAFFLED THAN EVER. THE CRUX... IS THAT THE VAST MAJORITY OF THE MASS OF THE UNIVERSE SEEMS TO BE MISSING."

WILLIAM BROAD, 1984

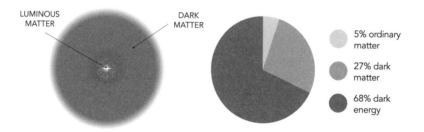

The universe is an immense place, containing billions of galaxies, many with billions of stars. Yet there are strong indications that the contents of the universe that we know about amount to only about 5 per cent of what makes it up.

Around 27 per cent is thought to be dark matter. This is matter that is invisible – it doesn't interact electromagnetically – but that has a gravitational effect. We are only aware of its existence because of its influence on the motion of galaxies and clusters of galaxies. Typically, they are spinning faster than they ought to be capable of with the amount of matter astronomers estimate they contain. There shouldn't be enough gravitational pull to hold them together, so we assume there is extra matter there that we can't see.

Dark energy accounts for the remaining 68 per cent of the universe. This is a mysterious energy causing the expansion of the universe to accelerate. It is tiny on the local level, but over the whole, vast universe it represents so much energy that it is the equivalent of twice the mass of everything else, dark matter included.

Above left: Dark matter halo around a galaxy.
Above right: Division of matter/energy in the universe.

ORGINS

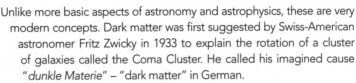

Unlike more basic aspects of astronomy and astrophysics, these are very modern concepts. Dark matter was first suggested by Swiss-American astronomer Fritz Zwicky in 1933 to explain the rotation of a cluster of galaxies called the Coma Cluster. He called his imagined cause "*dunkle Materie*" – "dark matter" in German.

The idea was largely ignored until the 1970s when American astronomer Vera Rubin, working with the instrument designer and astronomer Kent Ford, observed that some galaxies were not rotating as would be expected. A galaxy isn't a solid disk: all the stars do not rotate at a fixed rate with respect to each other, as would happen if they were directly connected. The expectation was that the rotation rate would shoot up from a relatively low rate at the centre, then gradually drop off at greater distances away. In reality, though, the rotation rate was relatively flat. Stars far away from the centre were rotating at a similar rate to those near the hub.

Rubin deduced that there was a hollow sphere of undetected matter around the outside of each galaxy, given the rather misleading name of a halo. Such a distribution would be unlikely for normal matter, but would be sensible for matter that only interacted gravitationally.

DARK ENERGY

Dark energy is an even more recent discovery. It had been known since Hubble measured the redshifts of galaxies that the universe was expanding. It seemed inevitable that over time this expansion would have slowed down as the gravitational attraction between the matter in the universe acted as a brake. In 1997, two groups studied data on distant supernovas which could be used to measure distances far greater than standard candles that had been used before. As this enabled them to look a number of billion years back in time, they expected to discover the rate at which the expansion was decelerating. Instead, to their surprise, they found that it was accelerating.

It took a number of years to assemble more confirming data, but the existence of dark energy, driving the expansion to accelerate, is now widely accepted, though there is no generally accepted explanation for it.

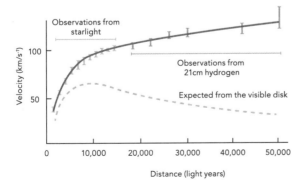

Top: Fritz Zwicky (1898–1974).
Above: Vera Rubin (1928–2016).
Left: Rotation curve showing deviation of speed in stars in a galaxy from expectation.

KEY **THEORIES** AND **EVIDENCE**

MISSING MATTER, DETECTION ATTEMPTS AND UNEXPECTED ACCELERATION

With plenty of evidence that something was causing galaxies and galactic clusters to rotate in a way that didn't quite fit with the general theory of relativity, much speculation went into the nature of the missing matter. It needed to be made up of particles that did not interact electromagnetically but had mass.

The first candidate was the neutrino, a neutrally charged, low-mass particle that is produced in vast quantities in nuclear reactions. Trillions of neutrinos from the Sun pass through you every second without being noticed – they score well on the not-interacting-electromagnetically scale. However, their mass is very low (they were originally thought to have no mass at all, like photons), and they move extremely fast. No workable model using neutrinos as dark matter has made it feasible that they would be captured in the vast halos thought to exist around galaxies.

WIMPS AND MACHOS

The alternatives have broadly been divided into two types – WIMPs (Weakly Interacting Massive Particles) and MACHOs (Massive Compact Halo Objects). MACHOs would be ordinary matter that we can't see, such as dust and black holes. However, there are severe problems in coming up with sufficient mass in MACHOs. Cosmologists have calculated how much ordinary matter should have been available in the universe, and it is significantly less than the matter plus dark matter total. And, as we have seen, ordinary matter is unlikely to form spherical halos.

WIMPs would have to be a relatively massive particle that sits outside the standard model of particle physics. But there is a problem here too. Theoreticians suggest that WIMPs should have mass that is accessible to particle accelerators such as the Large Hadron Collider – but no such particles have ever been detected in a collider.

Many other experiments have been set up to detect dark matter particles directly. Although they don't interact with ordinary matter through electromagnetism, there is still a small possibility of a direct collision with an atomic nucleus. (The same is the case with neutrinos, which are regularly detected.) Special detectors are set up deep underground to minimize interference from ordinary matter particles. They look out for tiny flashes of light – the energy released in a collision. Despite years of searching, not a single confirmed sighting of a dark matter particle has been made.

There have also been problems along the way with dark energy, but here the initial issue was being able to measure distances to galaxies that were billions of light years distant. This was too far away for the standard candles of the time, which were variable stars. But two competing teams, one of astronomers, the other physicists, managed to show that a particular type of supernova – a type Ia supernova – could be used. They couldn't just compare brightness. A supernova is an explosion that flares up to an intense brightness, then gradually dims. But they could compare the plot of the brightness as it faded away and discovered (on relatively close supernovas where other distance measures could be used) that the shape of the curve was distinctive for a particular brightness. This meant that the brightness curves could be used as a distance measure.

"THERE HAVE ALSO BEEN PROBLEMS ALONG THE WAY WITH DARK ENERGY, BUT HERE THE INITIAL ISSUE WAS BEING ABLE TO MEASURE DISTANCES TO GALAXIES THAT WERE BILLIONS OF LIGHT YEARS DISTANT."

MEASURING THE UNIVERSE

One reason this kind of measurement only became possible in the 1990s is the change in the way that telescopes were used. Historically people made direct observations, looking through a telescope. From the second half of the nineteenth century, though, photography began to be adopted. This hugely improved astronomy as exposures could be made over a length of time, picking up relatively dull stars, while the whole field of view could be examined at leisure. However, the breakthrough that made the supernova measurements possible was electronic sensors, like those in digital cameras. This meant the *changing* light level could be measured, not just a snapshot image.

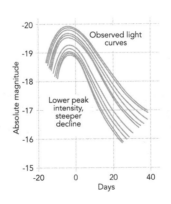

Opposite: The Coma Cluster of galaxies, Zwicky's original evidence for dark matter.
Above top: Large modern telescopes use electronic sensors rather than direct observation.
Above middle: DEAP-3600 dark matter detector located in Ontario, Canada.
Above bottom: Curves showing the change in supernova intensity help identify their actual brightness.

CRITICS

"EINSTEIN REMARKED TO ME MANY YEARS AGO THAT THE COSMIC REPULSION IDEA WAS THE BIGGEST BLUNDER HE HAD MADE IN HIS ENTIRE LIFE."

GEORGE GAMOW, 1956

Although dark matter remains the most widely accepted explanation for galactic rotation speeds, the existence of dark matter is only indirectly inferred and, to date, every attempt to detect it directly has failed. There is an alternative view that the theory of gravitation has to be modified when dealing with bodies on the scale of a galaxy.

The first theory to support this was modified Newtonian dynamics (or MOND), devised in 1983 by Romanian-Israeli astrophysicist Mordehai Milgrom. MOND has since been joined by a number of variants dealing with specific issues. Supporters of dark matter point to specific examples, such as a strangely shaped galactic cluster known as the Bullet Cluster, which don't fit well with the MOND hypothesis. However, what they rarely point out is that many other galaxies are a better match to the predictions of MOND than dark matter.

NO DARK MATTER?

One American mathematician, Donald Saari, has gone even further, suggesting that the analysis of the effect of ordinary matter in as complex a structure as a galaxy has been miscalculated and there is no need for dark matter at all.

While the existence of dark energy is widely accepted, the measurements involved are very difficult and could still prove inaccurate. However, disputes tend to be over the cause. One way to produce the effect of dark energy is to add in a cosmological constant – such as that dismissed by Einstein – but with a different value. This doesn't explain what causes the effect – at any one time there have been at least 50 different theories, but as yet there is no way to distinguish a successful one.

WHY IT **MATTERS**

Because dark matter and dark energy appear to form so much of the content of the universe, getting a better handle on what they involve is crucial to understanding how the universe is put together. Dark matter either challenges the standard model of particle physics, or aspects of the general theory of relativity. It is also a rare example where there are two opposing factions, for and against the existence of dark matter, each with strong scientific validity. Scientists are not used to being in this position: many supporters of dark matter speak as if there is no doubt about its existence.

Meanwhile, dark energy stretches the limits of our technology in getting accurate measurements of vast distances and equally has so far proved impossible to explain with a widely accepted theory. Although some might represent dark matter and dark energy as a failure of modern science – after all, we don't know what 95 per cent of the universe is – it is more reasonable to say that it presents a wonderful challenge for young scientists entering the field. At the end of the nineteenth century, some were announcing the end of science, with just a few details to be sorted out. Quantum physics and relativity turned this picture on its head. Equally, dark matter and dark energy now make it clear that we are nowhere near a boring "we know it all" phase.

Opposite top: Mordehai Milgrom (b.1946).
Opposite bottom: The Bullet Cluster.
Above: Nebulae and galaxies – Hubble telescope.

FUTURE **DEVELOPMENTS**

Despite the consistent failure of detection experiments, new experiments for direct detection of dark matter continue to be constructed. At the same time, theoreticians are developing better theories of modified gravity and looking at more exotic options, such as a hybrid solution combining aspects of each theory. We can expect to see an eventual clarification of dark matter's existence.

In the dark energy field, much of the experimental side is dependent on new, better telescopes, able to find more distant supernovas and to provide more clarity on exactly how far back in time we are looking in imaging the most distant galaxies. Theoreticians currently dispute whether dark energy is a constant effect or one that varies with time.

"AS IS OFTEN THE CASE WITH THE MORE EXTREME ASPECTS OF COSMOLOGY, IT COULD TAKE A LONG TIME TO MOVE FROM SPECULATION TO ANYTHING WITH A CHANCE OF BEING WIDELY ACCEPTED."

Some of the theories on how dark energy works have already been dismissed due to contradictory data. Other theories keep being added, though, some of which show no sign of being testable. As is often the case with the more extreme aspects of cosmology, it could take a long time to move from speculation to anything with a chance of being widely accepted. It may never happen, but it may be that getting to a widely accepted theory for the cause of dark energy will require the same kind of upheaval as getting to a theory of quantum gravity – the two could even be linked.

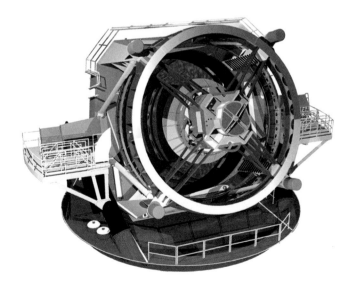

Above: The Vera C. Rubin Observatory, operational from 2022, will search for evidence of dark matter and dark energy.

THE **ESSENTIAL** SUMMARY

ORIGINS	KEY THEORIES AND EVIDENCE	CRITICS	WHY IT MATTERS	FUTURE DEVELOPMENTS
1933 Fritz Zwicky notices that the Coma Cluster rotated faster than expected, suggesting the existence of extra, invisible, "dark matter".	Around **5% of the universe is ordinary matter**, with around 27% dark matter and 68% dark energy.	Since Mordehai Milgrom suggested apparent dark matter could be a small error in general relativity when dealing with galaxies, **modified gravity theories** have suggested that dark matter might not exist.	Dark matter and dark energy appear to amount to **95% of the universe** which we currently don't understand.	**New detection experiments** for dark matter are being constructed.
Early 1970s Vera Rubin and Kent Ford observe unusual curves of the rotation speeds of stars in galaxies, suggesting that most galaxies have dark matter halos.	Various particles, including neutrinos, MACHOs and WIMPs, have been suggested as **dark matter particles**. None has been directly detected.	Mathematician Donald Saari has suggested that dark matter effects are **a calculation error** in the way stars influence each other gravitationally.	The debate over dark matter is of interest to **sociology of science professionals** as it is rare that two strongly opposing factions have similarly strong theories.	Theoreticians are developing better theories of **modified gravity** to show that dark matter isn't necessary.
1997 Two groups study distant galaxies and discover that the rate of the expansion of the universe is accelerating, driven by a mysterious "dark energy".	Dark energy was discovered when **supernovas** were used to measure the distance to distant galaxies.	Einstein called the cosmological constant his **greatest blunder** – but it can be used to represent dark energy.	Dark matter and dark energy present a **fascinating challenge** for future scientists: science has not yet become boring because we "know everything".	New telescopes are making **more accurate measurements** of the range of various distant galaxies, making it easier to clarify the scale of dark energy.
	It was not possible to use supernovas as standard candles until **electronic sensors** were used on telescopes, measuring the changing intensity of the supernova over time.	There are at least **50 different theories** attempting to explain dark energy.		**Some theories of dark energy have been dismissed**, but others continue to be added.

THE MULTIVERSE AND OTHER MYSTERIES

THE **ESSENTIAL** IDEA

"ONE COULD SAY: 'THE BOUNDARY CONDITION OF THE UNIVERSE IS THAT IT HAS NO BOUNDARY.'"

STEPHEN HAWKING, 1988

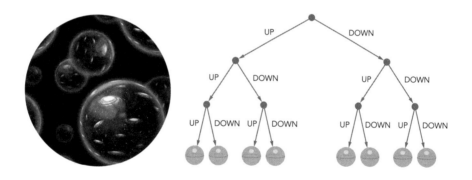

Some of the most speculative but fascinating ideas in cosmology and physics relate to the universe being something more than a single, unified physical entity. The Big Bang gives us a picture of a universe that starts from an infinitesimally small point and grows to its current size. But there could be far more to the universe than this.

There is nothing in our knowledge of the universe that says that it is limited to that Big Bang picture. A common concept suggests that our universe is just a small part of a larger "multiverse" – that the true universe consists of many smaller expanding universes, like a series of bubbles expanding on the surface of a sheet of water (but in three dimensions). Each bubble is independent and knows nothing of the other universes.

One of the interpretations of quantum physics is sometimes confused with the multiverse hypothesis. This "many worlds" interpretation also involves multiple universes, but here each is effectively an alternate version of its neighbour universes, splitting into two each time there is an option of two different outcomes from a quantum event.

ORIGINS

Although there have been historical references to "many worlds" which seem to pre-echo multiverse-like concepts, these only refer to there being more than one habitable planet in the universe.

One of the concepts to inspire the idea of the multiverse was the anthropic principle. In its weakest form, this is a straightforward logical deduction. It says that because human beings exist to observe it, the universe must work in such a way that human beings *can* exist. Although this appears trivial, it has been used to deduce at least one piece of scientific evidence by astrophysicist Fred Hoyle.

BUILT FOR HUMANS?

Much more controversial is the strong anthropic principle, which says that there are so many coincidences required to have the universe the way it is for humans to exist – many physical constants, for example, would only have to be slightly different for life not to exist – that there must be a mechanism that makes such an improbable universe possible. Supporters of the strong anthropic principle say that because a multiverse could contain universes with a whole range of different constants of nature, it would make it far more likely that at least one universe allowed humans to exist – hence our being here.

The many worlds hypothesis was introduced in American physicist Hugh Everett's PhD thesis in 1957 to get around a problem in quantum physics where it isn't clear how an initial set of probabilities "collapses" to a specific outcome when quantum particles interact. The many worlds interpretation gets around this problem by never having the probabilities collapse. Instead, all possible outcomes occur in different universes – but we only witness one outcome because we live along one path through the different worlds.

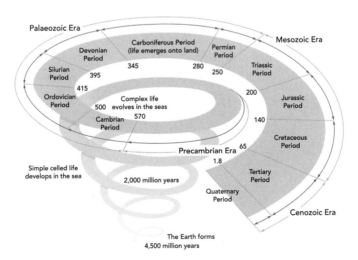

Opposite left: A multiverse could consist of many bubble universes, each with its own Big Bang.
Opposite right: The many worlds interpretation has universes splitting with each possible outcome of quantum events.
Above: The development of life on Earth, dependent on many physical constants being close to actual values.

KEY **THEORIES** AND **EVIDENCE**

BUBBLES, BRANES AND PARALLEL REALITIES

Our knowledge of the universe is limited to the distance light can travel in the time that our part of the potentially wider universe has existed. With the expansion of the universe, we know of the existence of a region about 90 billion light years across. Most cosmologists and physicists agree that we don't know what is beyond that horizon – it could be that there is a wider universe that is infinite in expanse.

This could be a wholly consistent environment, with the same natural laws and constants of nature that we experience. However, it is also possible that there have been many Big Bangs within the scope of a wider multiverse, each providing an expanding universe. If that is the case, each universe could be expanding (bearing in mind that our universe is not expanding *into* space – it is space itself that is expanding).

While it is possible that each universe has the same natural laws and constants of nature – that these are imposed on the wider universe – it is also possible that each universe has its own natural laws and constants.

UNIVERSAL LOTTERY

The way that the anthropic principle is used to argue for a multiverse can be illustrated using lottery tickets. Imagine that having a universe with the right laws and constants is like having a winning lottery ticket. Let's say there are a million tickets, only one of which will win. If there is a single universe, it is as if only one ticket of the million is issued, and it happens to be the winning ticket. It's very unlikely. The multiverse is like every ticket being issued – so *someone* will win. It's still very unlikely any particular universe will have survivable laws and constants, but one will. And because we are here to observe it, it happens to be our universe.

Above: The argument for the multiverse echoes the potential to win a lottery.

A variant on the multiverse could emerge if string theory is correct, which allows for extra dimensions of space we can't observe. If this were the case, our universe could be a three-dimensional object (known as a brane) floating in that multidimensional space… and there could be many more such branes, separate from our own.

Meanwhile, the many worlds hypothesis gives us something closer to a series of parallel realities, each almost indistinguishably different from the next, where a quantum choice has been made. So, for example, if a quantum particle could be spin up or spin down, and we usually say it takes on one of these values when it interacts with its surroundings, in the many worlds interpretation, the universe splits into two – one where the particle has spin up and one spin down.

EVIDENCE

There is no evidence whatsoever for a multiverse, and it is likely there never can be, as information from one universe in a multiverse cannot get to another universe. The concept remains pure speculation. A few scientists have argued that there may be special measurements that could in principle (though not in practice) be made which could distinguish between a multiverse and a single universe, but their suggestions are not widely supported.

The only argument for the multiverse is the idea that our current universe is "very unlikely". So, for example, the strong nuclear force which holds quarks together and binds the atomic nucleus would only have to be around 2 per cent stronger than it is in our universe and stars would no longer work. But, of course, unlikely things do happen and we would not be here to observe them happening if the universe did not support life.

There has also been at least one suggestion of a mechanism allowing for detection of the impact of the many worlds interpretation, but it is impossible to put into action and the idea remains without supporting evidence.

Above: The many worlds hypothesis gives a picture of a series of parallel realities.

CRITICS

The medieval concept of Ockham's razor – that, in effect, we should not make explanations more complex than they need to be – is often used against the many worlds interpretation. It seems extravagant to argue that there are a near-infinite set of universes, a whole new universe spawned every time a tiny quantum particle can have more than one outcome. The majority of physicists do not accept this interpretation.

As for a cosmological multiverse, some argue that this is not science. Science requires that a theory should be falsifiable. Many theories can't be proved to be true, but if there is a way to disprove them, and plenty of available evidence fails to do this, we can give support to the theory. But if there is no mechanism to show that a theory is false, it cannot truly be considered scientific. For example, take the theory that there is a giant invisible dragon sitting in the middle of Washington, DC – this dragon cannot be detected with any instrument. I can hypothesize that it's there, but without any ability to disprove its existence, that hypothesis is not science; it is fantasy. Many argue that the multiverse falls into a similar category.

WHY IT **MATTERS**

At best, theories like the multiverse can be described as mind-expanding or fun. Because they don't have any evidence to support them, they cannot be used for any direct scientific purpose, nor can we expect any practical applications.

The fact remains that theories like the multiverse capture some people's imaginations. It is the speculative, dramatic, edgy aspects of science from time travel to conscious robots that tend to spark an interest. There is little doubt that some individuals have developed an interest in science and have become important scientists because of the fascination of this kind of mind-bending idea.

It is also true, however, that such ideas have a downside. Individuals can spend their careers engaged in speculation that never leads to an outcome, while having academics working on concepts that even other scientists regard as pointless or not even science at all can give the public and funding bodies a negative view of those involved. This is something that needs to be considered, especially when politicians are regularly querying the value of experts.

Opposite top: Ockham's razor suggests we should keep hypotheses as simple as possible.
Opposite bottom: A hypothesis without the means to test it is comparable to suggesting there is an undetectable dragon.
Above: Dramatic, speculative aspects of science can attract individuals to the field.

FUTURE **DEVELOPMENTS**

Some cosmologists believe that the cosmic microwave background radiation could show evidence of collisions between our universe and others. No evidence that could be interpreted in this way has yet been found, but it is possible in the future, although it is hard to imagine, given the indirect nature of the evidence, that such data could not have other reasons for existing. Such findings would certainly not make a multiverse definitively real.

Similarly, the English physicist Roger Penrose has suggested that an artificial intelligence with a quantum computer for a brain could somehow gain information from across local parts of the many worlds multiverse, making it capable of distinguishing the many worlds interpretation from other quantum interpretations, but given that such a device is beyond our current comprehension and that few others agree, this is unlikely to deliver an answer.

There will no doubt be plenty more speculation on the existence of different forms of multiverse and the implications they would have – it is an enjoyable occupation for some, even if it remains more a matter of science fiction than science fact.

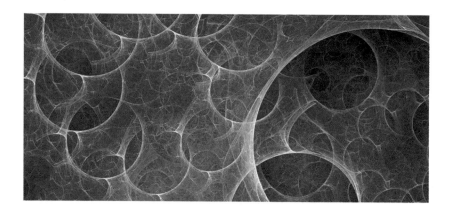

Top: Roger Penrose (b.1931).
Above: Speculation on different forms of multiverse will no doubt continue.

THE **ESSENTIAL** SUMMARY

ORIGINS	KEY THEORIES AND EVIDENCE	CRITICS	WHY IT MATTERS	FUTURE DEVELOPMENTS
Historical references to there being "many worlds" refer to other planets, not a true multiverse. **The anthropic principle** that the universe has to be set up a particular way for us to exist has been used to argue that with such an unlikely universe it ought to be one of many possible universes. **1957** Hugh Everett argues that to avoid the need for quantum probabilities to collapse into a single observed value it would be better if the universe split into a different universe for each possible outcome.	We don't know what lies beyond the **90-billion-light-year diameter** of the observable universe. There could have been **many Big Bangs** in a wider multiverse, each forming its own separate expanding universe. With many universes, **each could have its own physical laws** – our own universe may be unlikely, but its particular combination could occur at least once, rather like a winning lottery ticket. In a string theory universe, **three-dimensional branes** could float in a multidimensional multiverse, each brane being a complete universe. In a **many worlds universe**, when a particle could be spin up or spin down we get two universes, one for each outcome.	**Ockham's razor** suggests that we shouldn't make explanations more complex than they need to be – and the many worlds interpretation seems particularly extravagant. Many argue that the multiverse is not science at all as it **cannot be falsified**. Like a **giant undetectable dragon**, without evidence or a way of disproving the idea of a multiverse, some argue it is little more than fantasy.	Such theories can be **mind-expanding** or fun. Extravagant theories such as the multiverse and many worlds **capture people's imagination** and lead some individuals into science. However, individuals can waste careers on **speculative work** and disagreements about such work can lead outsiders to doubt the value of science.	Some believe that the **cosmic microwave background** could contain evidence of **universe collisions**, but none has been found and it would be hard to see how such data could be definitive. Roger Penrose has suggested that an **AI with a quantum brain** could detect interaction between many worlds universes – but others doubt this and such technology is beyond current capability. There will be **more speculation** in this field, which many will enjoy.

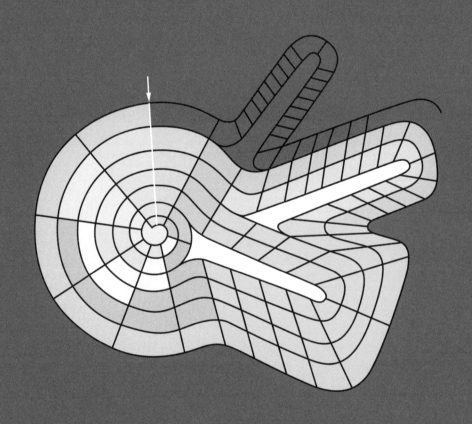

02

CHEMISTRY

ATOMIC STRUCTURE

THE **ESSENTIAL** IDEA

"I DON'T THINK THE ROYAL SOCIETY OF CHEMISTRY HAS AN OFFICIAL DEFINITION [OF CHEMISTRY], BUT I THINK IT'S ONE OF THOSE THINGS THAT 'YOU KNOW IT WHEN YOU SEE IT'."

RSC SOURCE, 2017

Chemistry tends to be a less visible science than physics or biology. One reason may be that it is difficult to pin down exactly what chemistry *is*. At one point it might have been defined as the science of the elements and how they combine – but a more precise definition might be the science of atomic structure, because almost all of chemistry is either concerned with the internal components of an atom or the distribution of electrons on the outside of an atom.

As we have seen, atoms were defined as being the "uncuttable" limit of matter, something with no internal configuration, but we now know that they have a crucially important structure. From the viewpoint of chemistry, the physicists' realization that the main components of the nucleus have their own substructure is irrelevant. All that matters to chemistry is the number of protons and neutrons in the nucleus and the distribution of electrons around it.

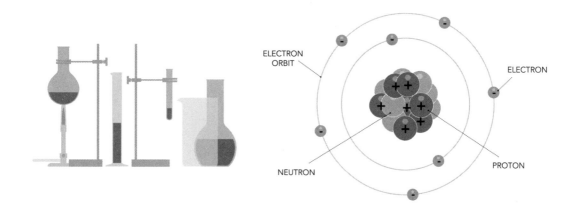

Above left: Chemistry was once seen as the science of the elements and how they combine.
Above right: More precisely, chemistry is the science of atomic structure.

ORIGINS

The English scientist John Dalton was the first to make a clear attempt to identify the structure of materials based on molecules, made up of combinations of atoms, and to assign relative weights to the different atoms. In 1803 he listed weights for a number of atoms, based on hydrogen, the lightest, having a weight of 1. He also deduced the proportions of the elements in a few compounds. Dalton had poor equipment even for the time and got some of the numbers wrong, yet it was a major step forward.

At the time, many did not accept that atoms existed, but by the end of the nineteenth century, as acceptance grew, it became clear that atoms were not the smallest components of matter. In 1897, English physicist Joseph John Thomson (universally known as J. J.) identified the electron as a separate particle. He would develop the "plum pudding" model of the atom comprising a number of negatively charged electrons, suspended in a positively charged massless region. As the lightest atom, hydrogen, was more than a thousand times heavier than an electron, this meant that Thomson suggested hydrogen contained over a thousand electrons (we now know it has one).

THE RUTHERFORD REVOLUTION

In 1909, Thomson's model was shattered. Working in Manchester, the New Zealand-born physicist Ernest Rutherford directed his German and English assistants, Hans Geiger and Ernest Marsden, in an experiment that transformed the view of the atom's structure. In the experiment, alpha particles (positively charged helium nuclei) were fired at a piece of gold foil. If the plum pudding model had held, it would be expected that particles would be slightly diverted by the diffuse charge. Instead, most went straight through, but some bounced back, showing that there was a heavy mass concentrated at the centre of the gold atoms. Taking the name from biology, Rutherford called this the atom's nucleus.

The remaining steps to the understanding of atomic structure would come with the realization of the quantum nature of the electron orbitals started by Danish physicist Niels Bohr in 1913, the detection of protons in the nucleus by Rutherford in 1917 and the 1932 discovery of the neutron by English physicist James Chadwick.

Above right: J. J. Thomson (1856–1940).
Above left: Dalton's table of chemical elements.

KEY **THEORIES** AND **EVIDENCE**

NUCLEI, ORBITALS AND VALENCE

"CHEMICAL ANALYSIS AND SYNTHESIS GO NO FARTHER THAN TO THE SEPARATION OF PARTICLES ONE FROM ANOTHER, AND TO THEIR REUNION. NO NEW CREATION OR DESTRUCTION OF MATTER IS WITHIN CHEMICAL AGENCY. WE MIGHT AS WELL ATTEMPT TO INTRODUCE A NEW PLANET INTO THE SOLAR SYSTEM, OR TO ANNIHILATE ONE ALREADY IN EXISTENCE, AS TO CREATE OR DESTROY A PARTICLE OF HYDROGEN."

JOHN DALTON, 1808

Central to chemistry is the idea that matter, from the simplest hydrogen gas to the most complex living organic material, is composed of molecules, themselves made up of atoms. There are 94 different types of atom known in nature (another 24 have been created, but are too unstable to exist naturally), each making up a specific chemical element. Some of these have been known since ancient times, while others were discovered far more recently.

The chemical properties of an element are dependent on its atomic structure. Each atom consists of a nucleus, containing the vast majority of its mass, made up of positively charged protons and electrically neutral neutrons, and a surrounding collection of low-mass, negatively charged electrons. The "atomic number" assigned to each element is the number of protons in the nucleus. So, for example, hydrogen, the lightest element, has one proton; helium, the next heaviest, has two.

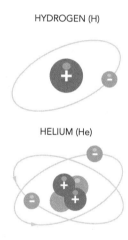

HYDROGEN (H)

HELIUM (He)

ATOMIC STRUCTURE

The number of neutrons in the nucleus of a particular element can vary. Atoms with the same number of protons but differing numbers of neutrons are known as isotopes. Chemically, isotopes behave identically, but some are unstable and can decay radioactively to form different elements. Take, for example, the carbon atom. This has six protons, but can have five, six, seven or eight neutrons. Each isotope is identified by adding together the number of protons and neutrons.

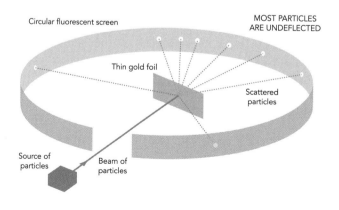

Circular fluorescent screen

MOST PARTICLES ARE UNDEFLECTED

Thin gold foil

Scattered particles

Source of particles

Beam of particles

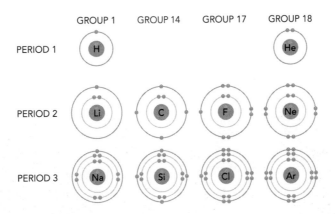

The most common isotope with six neutrons is known as carbon-12, sometimes written as ^{12}C, where C is the chemical symbol for carbon.

Around the nucleus, electrons are distributed in three-dimensional regions of probability called orbitals. The electrons group together in "shells" which are occupied by electrons with different properties, such as spin. Each shell has a maximum capacity of electrons. The number of electrons present in the outermost shell define how the element will behave chemically, as the interactions of different elements are caused either by electrons being removed or added to the outer shell, or by sharing of outer-shell electrons between atoms. Stable compounds tend to form when the combined electrons in the outer shells result in the outermost shell being full. The number of electrons in the outer shell (or the number of gaps to fill it) is referred to as the atom's valence, which dictates how it can react with other atoms.

EVIDENCE

The existence of the structure of atoms was first discovered by experiments such as the Rutherford gold foil experiment. Protons were discovered relatively quickly because a hydrogen ion (the atom with its electron missing), which is common in nature, is just a proton, however it took some time for neutrons to be detected as this required a nuclear reaction where the nucleus of an unstable atom broke down, emitting neutrons in the process.

Isotopes were discovered by English chemist Frederick Soddy, working on radioactivity in Canada with Ernest Rutherford. Their existence is suggested by the way that the weight of some atoms is not a round multiple of the hydrogen atom – for example, chlorine has the atomic weight 35.45, made up of a mix of isotopes with weight 35 and 37.

More detail of the atomic nucleus came from further studies of radioactivity, while the electronic orbitals and shells were initially derived mathematically as quantum physics was developed, giving a better understanding of how different properties of the electron would influence their ability to share space. At the heart of this is the Pauli exclusion principle, which says that two electrons can't be in the same quantum state in the same atom – at least one of their properties must be different. A small number of orbitals have now been experimentally mapped out using special electron microscopes.

Opposite top: Atomic structure of the simplest elements.
Opposite bottom: Rutherford's experiment establishing the existence of the atomic nucleus.
Above: Electrons in the shells of a range of atoms.

CRITICS

Initially there was considerable opposition to the existence of atoms. William Thomson, Lord Kelvin, for example, was of the opinion that what were described as atoms were in reality vortexes in the ether, the medium that was thought to fill all of space.

There was some fair criticism of Dalton's early work – as well as getting atomic weights wrong, he was of the opinion that atoms would form the simplest possible combination. So, for example, he thought that water was HO, with one hydrogen and one oxygen atom, rather than the correct H_2O with two hydrogen atoms for every oxygen atom. It was rightly pointed out that this was an arbitrary imposition rather than anything suggested by evidence.

"LORD KELVIN, FOR EXAMPLE, WAS OF THE OPINION THAT WHAT WERE DESCRIBED AS ATOMS WERE IN REALITY VORTEXES IN THE ETHER"

The critics who suggested that atoms didn't exist were generally silenced by Einstein's work on Brownian motion. This was the discovery by Scottish botanist Robert Brown that pollen grains jumped around when suspended in water. He first thought this was due to some kind of life force in the pollen, but discovered that it also occurred with lifeless materials. Einstein showed mathematically how the motion would be caused by water molecules, and French physicist Jean Perrin went on to use Einstein's work experimentally to define the mass and size of atoms.

While there would inevitably be some criticism of the new ideas of the quantum physicists, the combination of the development of theory and experiment, particularly with radioactive materials, in the 1920s and 1930s quickly made support for the atomic structure universal.

Above left: William Thomson, Lord Kelvin (1824–1907).
Above right: Jean Perrin (1870–1942).
Opposite: Atomic structure is essential to understand everything from chemistry to nuclear fusion.

WHY IT **MATTERS**

The structure of the atom may seem like knowledge for the sake of it. Just as is the case with, say, our understanding of cosmology, it expands our appreciation of the world around us and is worthwhile it its own right. However, unlike cosmology, understanding the structure of the atom has massive practical benefits.

As all chemical reactions are dependent on atomic structure (and nuclear reactions on nuclear structure), gaining a good understanding of how atoms are put together is the essential basis of chemistry. Whether we're trying to develop new drugs that target specific medical requirements or producing new materials – for example, to replace plastics derived from fossil fuels – a thorough knowledge of the structure of the atom is essential.

> "IT IS ALSO THROUGH AN UNDERSTANDING OF ATOMIC STRUCTURE THAT WE HAVE BEEN ABLE TO UNDERSTAND THE WAY THAT THE SUN WORKS"

It is also through an understanding of atomic structure that we have been able to understand the way that the Sun works and from this to look at mechanisms for using nuclear fusion to produce clean energy on Earth. Similarly, when attempting to improve the performance of solar cells, understanding the atomic processes used in nature through photosynthesis enables us to consider whole new opportunities for clean energy generation.

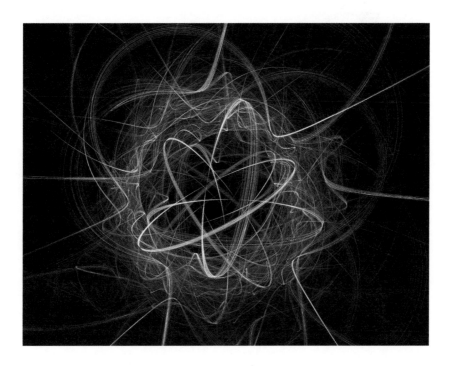

FUTURE **DEVELOPMENTS**

The basic understanding of atomic structure is unlikely to provide us with many future surprises. However, we are seeing significant developments in the way that we undertake chemistry, based on our knowledge of the atom. We have seen the production of previously non-existent chemical elements, all the way up to element 118, oganesson, only five atoms of which have so far been detected. More elements are likely to be added.

In the future, we are likely to see far more molecular assembly – effectively operating on individual molecules to cut apart and construct new structures and new materials. In biology this has already started, particularly with CRISPR (Clustered Regularly Interspaced Short Palindromic Repeats), a technique used to edit lengths of DNA, precisely cutting the molecule where required.

Longer term, it has been suggested that we could have nanotechnology that can operate at the level of atoms, using armies of "assemblers" to put atoms together in any desired combination, making possible the Star Trek style "replicators" used to generate food and drink, but capable of producing practically anything from basic raw materials. This is currently science fiction, but could in principle be possible.

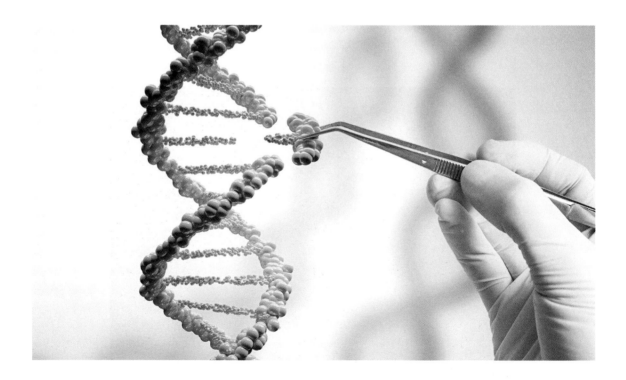

Above: The CRISPR technique can
be used to make precise edits of DNA.

THE **ESSENTIAL** SUMMARY

ORIGINS	KEY THEORIES AND EVIDENCE	CRITICS	WHY IT MATTERS	FUTURE DEVELOPMENTS
1803 John Dalton lists relative weights for a number of atoms, based on hydrogen as 1, and deduces the structure of some compounds. **1897** J. J. Thomson discovers the electron, the first subatomic particle, and suggests atoms have a "plum pudding" structure with electrons scattered through a positively charged matrix. **1909** Hans Geiger and Ernest Marsden, working for Ernest Rutherford, discover the atomic nucleus. **1913** Niels Bohr devises the quantum atomic structure. **1917** Rutherford discovers the proton. **1932** James Chadwick discovers the neutron.	**All matter** contains atoms or molecules, themselves made of atoms, in 94 natural elements. The **chemical properties** of an element depend entirely on its atomic structure. Each element has a fixed number of protons in the nucleus, while the outer electrons define how it will react. An element can have **differing numbers of neutrons** in the nucleus, making up variants known as isotopes. Electrons inhabit regions of space called **orbitals** and are grouped together in **shells**. The number of electrons in the outer shell – the **valence** – determines how the atom will react.	Initially there was considerable **resistance to the existence of atoms** – Lord Kelvin, for example, thought they were vortexes in the ether, not physical objects. Dalton's work was correctly criticized for **inaccuracy in atomic weights** and the structure of common molecules. Einstein's mathematical explanation of **Brownian motion** quelled most resistance to the idea of atoms. Although there was **some resistance to quantum theory**, support for the atomic structure rapidly became universal.	The structure of atoms is the kind of **essential knowledge** that is worthwhile in its own right. All **chemical reactions** are dependent on atomic structure. Whether **developing new drugs** or **designing new materials**, an understanding of atomic structure is essential. Through an understanding of atomic structure we have been able to devise **nuclear fusion** reactors and better understand **photosynthesis** to consider new approaches to solar cells.	We are likely to see more **new elements** added to the periodic table. Mechanisms such as **CRISPR** enable us to manipulate molecules: this will be expanded greatly in the future. There is speculation that we might in the future be able to build **nanotechnology assemblers** that could produce anything from a stock of atoms.

STATES OF MATTER

THE **ESSENTIAL** IDEA

"THERE ARE THREE DISTINCTIONS IN THE KINDS OF
BODIES, OR THREE STATES, WHICH HAVE MORE ESPECIALLY
CLAIMED THE ATTENTION OF PHYSICAL CHEMISTS; NAMELY,
THOSE WHICH ARE MARKED BY THE TERMS ELASTIC
FLUIDS [GASES], LIQUIDS AND SOLIDS."

JOHN DALTON, 1808

Chemical elements and molecules made from them are typically found in four different states. The familiar three have been known for centuries: gases, liquids and solids. In each case, the difference is in the way that the atoms or molecules move around and link together. In solids, they tend to be strongly linked, so that atoms vibrate, but do not move any significant distance through space. In a liquid, atoms (or molecules) can move individually, but still feel strong attraction towards others, keeping the liquid in a container under the force of gravity. In a gas, atoms or molecules move sufficiently quickly that they travel independently, though they will frequently change course due to collisions with other atoms and molecules.

The fourth state of matter is the most common one in the universe, because it is the state of matter in a star: this is a plasma. We also find plasmas wherever a material is strongly heated, for example in a flame. A plasma is like a gas, but rather than being made of atoms it is made of ions – atoms that have gained or lost electrons to become electrically charged.

Above: Flames contain a significant amount of plasma.
Opposite top: Joseph Priestley (1733–1804).
Opposite middle: William Crookes (1832–1919).
Opposite bottom: Irving Langmuir (1881–1957).

"CHEMICAL ELEMENTS AND MOLECULES MADE FROM THEM
ARE TYPICALLY FOUND IN FOUR DIFFERENT STATES."

ORIGINS

All four major states of matter exist in nature, but only solids and liquids, which can be seen and handled, have been recognized as different states of the same substances since antiquity. The original four elements devised by the fifth-century BC ancient Greek philosopher Empedocles coincidentally correspond well to the four states. Earth is a solid, water a liquid, air a gas and fire usually incorporates a plasma – but for the ancient Greeks these were four separate entities, rather than different states of the same substance.

Leaving aside the elements, the solid/liquid relationship was clear with substances such as water and metals that could be melted at high temperatures. It was harder to pin down the link to gases as they were usually invisible. While air was known to exist, due to the ability to feel wind and air resistance, it continued to be seen as a uniform, different kind of element.

THE DISCOVERY OF OXYGEN

It was only with the discovery of a mix of gases in the air by the likes of English philosopher Joseph Priestley, who discovered oxygen in 1774, and the realization that water was a compound of the two gases oxygen and hydrogen by English philosopher Henry Cavendish in 1781, that it was fully accepted that gases were simply another state of matter.

Plasma was a late addition. Glowing materials such as the Sun, flames, lightning and neon lights all contain plasma, and the specific material had been identified under the name "radiant matter" by English physicist William Crookes when studying early cathode ray tubes. But the idea of plasma being a fourth state of matter was only clarified in the 1920s by American chemist Irving Langmuir. Just as Rutherford had named the atomic nucleus after biology's cell nucleus, Langmuir named the newly identified state of matter after blood plasma, because of the analogous way blood plasma transports corpuscles and physical plasma transports charged particles.

KEY **THEORIES** AND **EVIDENCE**

STATES, PHASE CHANGES AND STRANGE MATTER

"A STRICT MATERIALIST BELIEVES THAT EVERYTHING
DEPENDS ON THE MOTION OF MATTER."

JAMES CLERK MAXWELL, 1868

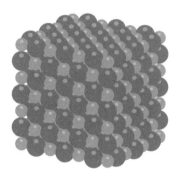

All matter is made up of atoms which, depending on their kinetic energy and degree of electromagnetic interaction, can be in a number of states. Solids generally do not flow, and have strong electromagnetic attraction between component atoms, holding them sufficiently firmly in place that they can only vibrate, not move at all freely. Some solids are electrically conductive as electrons can move through the lattice of atoms. A liquid can flow and take the shape of a container, but is sufficiently tightly packed that it cannot be compressed by a significant amount. Here there are still significant electromagnetic forces between atoms or molecules, but not sufficient to hold them in place.

WHAT IS A GAS?

A gas is also a fluid, but the atoms or molecules are sufficiently spaced out that there is relatively little electromagnetic attraction, and the substance can be compressed. Atoms or molecules will be moving sufficiently quickly that there is limited opportunity to attract each other. Gases (and liquids) are not usually electrically conductive unless they contain charged particles. Finally, a plasma is like a gas in structure, but instead of being made up of atoms or molecules, it is made up of ions: atoms that have typically lost one or more electrons (or in some cases have gained electrons), so they are electrically charged. Unlike a normal gas, a plasma is electrically conductive.

When a substance moves from one state to another it undergoes what is known as a phase change or transition. A solid can melt to form a liquid, or sublimate to transition directly to a gas. A liquid can evaporate to a gas, and a

Above: The tight-packed atoms of a solid in the sodium chloride crystal lattice.
Opposite left: Liquids can flow but cannot be significantly compressed.
Opposite right: Phase transitions between different states of matter.

gas can be ionized to form a plasma. Each transition has a reverse equivalent – freezing for the transition from liquid to solid, undergoing deposition from gas to solid, condensing for gas to liquid and deionizing from plasma to gas.

When a phase change occurs, it usually either requires energy to take place, or gives off energy in the process. This is traditionally described as the latent heat of the transition. So, for example, when a liquid is heated, when it reaches boiling point it will stay at that temperature as the energy provided goes to breaking bonds, transforming the liquid into its gaseous form.

Apart from the main states of matter there are a few strange states that are either intermediate or sufficiently unlike a conventional state to be considered something different. So, for example, the liquid crystals that feature in many display screens can typically flow like a liquid but have a structure that is more like a solid. There are a number of unusual low-temperature states when materials are close to absolute zero (−459.67°F or −273.15°C). These include superfluids, which flow without friction, and Bose–Einstein condensates. Here, the atoms in the material share a single quantum state, giving them unusual properties when interacting with other quantum particles such as photons.

"APART FROM THE MAIN STATES OF MATTER THERE ARE A FEW STRANGE STATES THAT ARE EITHER INTERMEDIATE OR SUFFICIENTLY UNLIKE A CONVENTIONAL STATE TO BE CONSIDERED SOMETHING DIFFERENT."

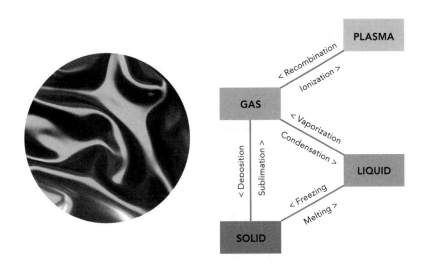

EVIDENCE

Most of the properties of states of matter and of transitions between them are the result of direct observation; in some cases they have been known for centuries in practical terms before it was understood what was happening at the theoretical level.

The more unusual states of matter were typically developed in the twentieth century. Liquid crystals were discovered and Bose–Einstein condensates were theorized in the 1920s, while superfluidity was observed in the 1930s, based on low-temperature work first carried out in 1911 by Dutch physicist Heike Kamerlingh Onnes.

CRITICS

Although it is usually clear what state a material is in, some materials have produced significant disputes, none more so than glass. For a long time, it was popularly suggested that glass was a very slow-flowing liquid. As evidence of this, it was pointed out that very old windowpanes are typically thicker at the bottom than they are at the top. This was assumed to be because, over the centuries, the glass slowly flowed down the pane. However, it was realized that old sheets of glass, made before modern processes, were usually thicker at one edge than the other, and glaziers would put the glass in place with the thicker part at the bottom for stability.

In reality, glass is an amorphous solid – it is not a crystal with a regular lattice, but rather a messy collection of atoms. Another example, demonstrating the unusual nature of an amorphous solid, is a gel. True slow-flowing liquids do exist – the best-known example is tar or pitch. An experiment featuring bitumen tar (the kind used in road-making) in a funnel has been running in Australia since 1927. As yet, just eight drops have fallen from it.

Above: The Australian tar drop experiment.
Opposite: Plasma in a nuclear fusion reactor.

WHY IT **MATTERS**

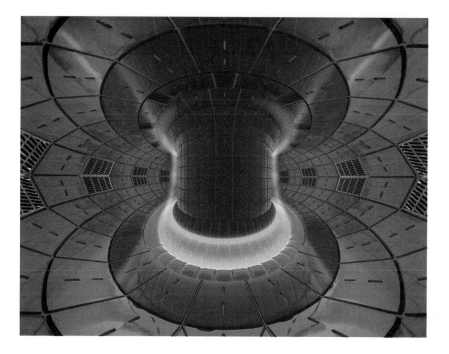

Understanding transitions between states of matter have been crucial for manufacturing ever since humans began to smelt metals, while the different states of water (ice, water and water vapour) are central to the existence of many species, as all living things require liquid water to survive. A good example of a long-standing industry that has needed to understand a particular state of matter is glassmaking, where the transition between liquid (required to shape the glass) and solid is critical.

> "FUSION, WHETHER IN THE SUN OR IN A REACTOR, INVOLVES PLASMAS, WHICH ARE TRICKY TO CONTROL AS THEY CAN SURGE AND JUMP AROUND AS IF THEY ARE ALIVE."

What used to be relatively obscure phases of matter – plasma and liquid crystals – have become essential for modern technology. Both are used in display devices, while understanding the behaviour of plasmas is central to our ability to use nuclear fusion for electricity generation. Fusion, whether in the Sun or in a reactor, involves plasmas, which are tricky to control as they can surge and jump around as if they are alive. Within a fusion reactor, the plasma has to be magnetically contained to avoid contact with ordinary matter, which would both cool the plasma and damage the containment vessel.

Overall, whenever we deal with materials of any kind, understanding the state of matter and any transitions it undergoes are essential to dealing properly with those materials.

FUTURE **DEVELOPMENTS**

As well as liquid crystals, superfluids and Bose–Einstein condensates, there are a number of other special states of matter, typically involving interaction with quantum particles, that can produce strange results. As yet these have limited applications, but it is likely they will be developed in the future. For example, in special materials, photons can be induced to interact with each other, producing what has been described as a photonic molecule, which is of interest to those developing future computers.

Closer to current reality are the many potential applications of a type of solid that is new to the twenty-first century: atom-thick materials. The leading example here is graphene, a one-atom-thick layer of graphite (carbon). Because it is so thin, it is subject to many quantum effects, making it, for example, both the best conductor and the strongest material known in response to stretching. Graphene and other atom-thick materials are likely to feature in the next generation of electronics, taking over from current silicon-based devices, which cannot be miniaturized much further.

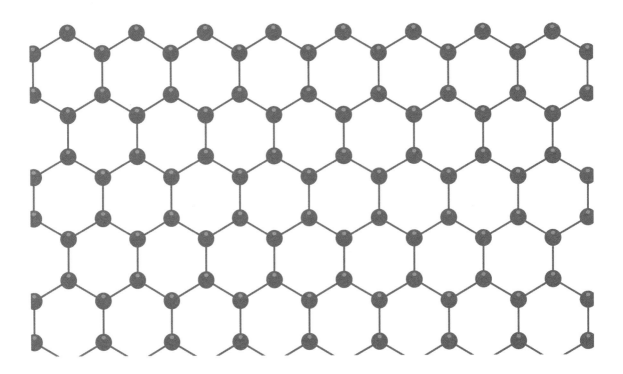

Above: The single-atom-thick lattice of graphene.

THE **ESSENTIAL** SUMMARY

ORIGINS	KEY THEORIES AND EVIDENCE	CRITICS	WHY IT MATTERS	FUTURE DEVELOPMENTS
5th century BC Greek philosopher Empedocles identifies four "elements" – earth, water, air and fire – which correspond to the four main states of matter: solid, liquid, gas and plasma. **1774** Joseph Priestley discovers oxygen, helping to identify gases as another state of matter. **1781** Henry Cavendish shows water is made of the gases hydrogen and oxygen. **1920** Irving Langmuir identifies plasma as a fourth state of matter.	In **solids** atoms are strongly attracted to each other. The matter does not flow but atoms can vibrate. A **liquid** can flow and take the shape of a container but there is still strong attraction between atoms. Atoms in a **gas** are free to move and the gas can be compressed. A **plasma** is structurally similar to a gas, but contains electrically charged ions rather than atoms and so can conduct electricity. When a substance **changes state** (a phase transition) it often involves heat being added or given off, known as the latent heat of the transition. There are a number of intermediate or strange states such as **liquid crystals, superfluids** and **Bose–Einstein condensates**.	For a long time, many thought that glass was a **liquid that flowed very slowly** – but the evidence of old sheets of **glass** proved to be due to early manufacturing issues. Glass is an amorphous solid. There are confusing, **slow-flowing liquids**, most notably tars such as **bitumen**. An experiment with bitumen flowing through a funnel has produced just eight drops since 1927.	Understanding **phase transitions** is essential to much **manufacturing**, while the phase transitions of water affect all **living things**. **Glassmaking** is a good example of an industry where understanding the **solid/liquid phase transition** is essential. Understanding **plasma** and **liquid crystals** have both become essential for modern technologies from **displays** to **fusion reactors**.	As well as liquid crystals, superfluids and Bose–Einstein condensates, other **strange quantum states of matter** have the potential for future applications. Of particular interest are ultra-thin, **atom-thick materials**. Best known is **graphene**, a single-atom-thick layer of carbon. This is ultra-strong and highly conductive and will provide many new opportunities in next-generation electronics.

THE PERIODIC TABLE

THE **ESSENTIAL** IDEA

"WHEN THE ELEMENTS ARE ARRANGED IN VERTICAL COLUMNS ACCORDING TO INCREASING ATOMIC WEIGHT, SO THAT THE HORIZONTAL LINES CONTAIN ANALOGOUS ELEMENTS AGAIN ACCORDING TO INCREASING ATOMIC WEIGHT, AN ARRANGEMENT RESULTS FROM WHICH SEVERAL GENERAL CONCLUSIONS CAN BE DRAWN."

DMITRY MENDELEEV, 1869

Because chemical reactions are dependent on the outer shell of electrons on an atom of an element, once a shell is filled and the next shell starts to fill, another element will come along with a similar set of outer electrons – leading to similar properties.

Although the periodic table of the elements was devised before chemists were aware of the reason why elements had similar properties – in fact at a time when there was still considerable dispute over whether atoms existed at all – it was arranged based on the similarities of the elements. As Mendeleev put it, these similar elements were "analogous": in the quote above, his horizontal lines correspond to columns in the modern table. The elements in a column would be known as a group and the position down the column as a period.

The periodic table became a powerful tool for understanding how the chemical elements relate to each other.

Reihen	Gruppe I. — R^2O	Gruppe II. — RO	Gruppe III. — R^2O^3	Gruppe IV. RH^4 RO^2	Gruppe V. RH^3 R^2O^5	Gruppe VI. RH^2 RO^3	Gruppe VII. RH R^2O^7	Gruppe VIII. — RO^4
1	H=1							
2	Li=7	Be=9,4	B=11	C=12	N=14	O=16	F=19	
3	Na=23	Mg=24	Al=27,3	Si=28	P=31	S=32	Cl=35,5	
4	K=39	Ca=40	—=44	Ti=48	V=51	Cr=52	Mn=55	Fe=56, Co=59, Ni=59, Cu=63.
5	(Cu=63)	Zn=65	—=68	—=72	As=75	Se=78	Br=80	
6	Rb=85	Sr=87	?Yt=88	Zr=90	Nb=94	Mo=96	—=100	Ru=104, Rh=104, Pd=106, Ag=108.
7	(Ag=108)	Cd=112	In=113	Sn=118	Sb=122	Te=125	J=127	
8	Cs=133	Ba=137	?Di=138	?Ce=140	—	—	—	— — —
9	(—)	—	—	—	—	—	—	
10	—	—	?Er=178	?La=180	Ta=182	W=184	—	Os=195, Ir=197, Pt=198, Au=199.
11	(Au=199)	Hg=200	Tl=204	Pb=207	Bi=208	—	—	— — —
12	—	—	—	Th=231	—	U=240	—	— — —

Above left: Mendeleev's early periodic table.
Above right: The number of electrons in an atom's outer shell determine its chemical properties.

ORIGINS

When John Dalton first identified relative atomic weights in 1803, he simply listed the elements (and a few compounds then thought to be elements) in order of increasing weight. However, it wasn't long before chemists were noting similarities in elements and structuring them together. For example, in 1829 the German chemist Johann Döbereiner thought he had identified some kind of internal structure in groups of three, which he called triads. For example, he noticed that chlorine, bromine and iodine all had similarities, as did lithium, sodium and potassium.

GROWING ELEMENTS

It took some time for reasonably accurate data to be derived for the growing list of elements, but by the 1860s, several chemists were extending the kind of structure that Döbereiner had noticed to take in more similarities. For example, the English chemist John Newlands in 1864 seems to have taken the same sort of inspiration Newton did with colours of the rainbow and thought that the elements could be listed in groups of seven, making them effectively octaves like musical notes, returning to a similar element at the start of the next octave.

This attempt did not go down particularly well – Newlands was trying to force a structure onto the elements that bore a limited relationship with reality. But a German contemporary, Lothar Meyer, produced a more convincing periodic table by not worrying about the numbers in each group, but rather putting elements in increasing weight depending on their valence – the ratio with which they would combine with other elements in compounds.

ENTER MENDELEEV

Meanwhile, the Russian chemist Dmitry Mendeleev was playing around with a structure for the elements, writing them on individual cards and rearranging them. His first table was published in 1869, and by 1871, Mendeleev had come up with a table that bears a resemblance to the modern table, including gaps for a number of elements that he felt ought to be in place but were yet to be discovered. He named them by using the substance in the table sitting above the gap with the prefix eka-, which is Sanskrit for the number one, attached. So, for example, below silicon Mendeleev said we should have ekasilicon, which occupied the slot we now assign to germanium.

Above top: Johann Döbereiner (1780–1849).
Above middle: Julius Meyer (1830–1895).
Above bottom: Dimitry Mendeleev (1834–1907).

KEY **THEORIES** AND **EVIDENCE**

GROUPS, PERIODS AND GAPS

"THE PERIODIC TABLE WAS INCREDIBLY BEAUTIFUL, THE MOST BEAUTIFUL THING I HAD EVER SEEN. I COULD NEVER ADEQUATELY ANALYSE WHAT I MEANT HERE BY BEAUTY – SIMPLICITY? COHERENCE? RHYTHM? INEVITABILITY? OR PERHAPS IT WAS THE SYMMETRY, THE COMPREHENSIVENESS OF EVERY ELEMENT FIRMLY LOCKED INTO ITS PLACE, WITH NO GAPS, NO EXCEPTIONS, EVERYTHING IMPLYING EVERYTHING ELSE."

OLIVER SACKS, 2001

At the heart of understanding the periodic table is the importance of the number of electrons in the outermost shell of an element. A lot of the confusion caused for early constructors of tables came from the fact that the shells do not all contain the same number of electrons. The innermost shell can hold a maximum of 2 electrons – 1 in the case of hydrogen and 2 for helium – so the first period (horizontal row of the table) contains just these two elements.

The next shell holds 8 electrons, giving the eight elements in the second period. Things get more complicated after that. The third shell can hold 18 electrons, but the outer shell only ever gets up to 8. After this, the fourth shell (which can hold 32 electrons) starts filling. Now the third shell is no longer the outermost shell it can go beyond 8 and heads up to 18 as the fourth shell heads to 8. As a result, the third period also has eight elements before leaping up to 18 in the next. The fourth period also runs to 18, taking the fourth shell to 18 and the fifth to 8.

This produces the typically jagged look of the periodic table. Strictly, the table should be 32 columns wide to accommodate all the current elements, but a better consistency was achieved (and a more compact table) by keeping the sixth and

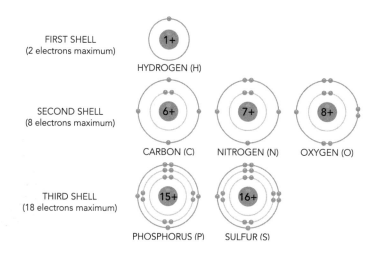

FIRST SHELL
(2 electrons maximum)

HYDROGEN (H)

SECOND SHELL
(8 electrons maximum)

CARBON (C) NITROGEN (N) OXYGEN (O)

THIRD SHELL
(18 electrons maximum)

PHOSPHORUS (P) SULFUR (S)

Above: The outer shells of atoms contain different numbers of electrons.
Opposite: The periodic table in its modern form.

seventh periods at 18, with a gap after the third element in each period where an extra 14 filler rows accommodating the elements known as the lanthanides and the actinides are positioned. These extra rows are named after the element that appears just before the point the filler row should be: lanthanum and actinium.

None of the elements above uranium (92) occur naturally on the Earth, and none above plutonium (94) have been detected in space. All the other elements were produced artificially in nuclear reactions. The heaviest element to date is oganesson (118), of which only five atoms have been produced. The half-life of these atoms (the time in which half would decay) is less than a millisecond. If elements beyond oganesson are produced a new period will be needed.

EVIDENCE

Initially the periodic table was built on a combination of factors. As well as being in order of increasing atomic weight, the elements were structured according to their chemical properties (for example, metals and non-metals cluster together) and valence. Because of the odd way that the electron structure develops, and the existence of isotopes, there are occasional anomalies. Mendeleev, for example, thought that the measured atomic weight of the element tellurium, thought to be 128, was incorrect as its position in the table based on its properties meant that it came before iodine, which had an atomic weight of 126. In fact, it is heavier than iodine (127.6 to iodine's 126.9), despite correctly appearing before it.

"ATOMIC WEIGHTS ARE CALCULATED FROM THE ENERGY REQUIRED TO ACCELERATE A CHARGED ION IN A PARTICLE ACCELERATOR, CORRECTED FOR THE MISSING ELECTRON(S)."

Much of the modern information for the table is derived from the electronic structure, which is deduced from a number of properties, notably the spectroscopic emissions when electrons jump up and down between orbitals. Atomic weights are calculated from the energy required to accelerate a charged ion in a particle accelerator, corrected for the missing electron(s).

CRITICS

Early attempts at producing a periodic table were criticized because the reasoning was arbitrary. Just as Newton had decided that there should be seven colours in a rainbow by worthless analogy with a musical octave, so Newlands took the same approach and was correctly opposed because there was no reason to draw such an analogy.

Mendeleev's approach was different, driven by logical connections, and it enabled him to make predictions that could be checked, such as giving a surprisingly accurate picture of the missing element germanium before it had been discovered.

"IT IS HARD FOR ANY OBJECTIVE OBSERVER TO SEE WHY MOISSAN WON OVER MENDELEEV"

Mendeleev was widely feted for his work, winning, for example, two of the Royal Society's most important medals: the Davy Medal in 1882 for a discovery in chemistry and the Copley Medal in 1905 for outstanding achievements in research. It seemed highly likely that Mendeleev would then win the Nobel Prize in Chemistry in 1906 for what surely must be one of the greatest pieces of work in the field. Instead it was awarded to the French chemist Henri Moissan "in recognition of the great services rendered by him in his investigation and isolation of the element fluorine, and for the adoption in the service of science of the electric furnace called after him". It is hard for any objective observer to see why Moissan won over Mendeleev, a result that seems to have been down to politics on behalf of the Nobel committee. As Mendeleev died in 1907 and the Nobel Prize is not awarded posthumously, he would not get another chance.

Above left: The Copley medal's recipients included
the pioneer of antiseptic surgery, Joseph Lister.
Above right: Henri Moissan (1852–1907).
Opposite top: The iconic periodic table tiles.
Opposite bottom: Regions of the table showing metals (left),
metalloids (diagonal), non-metals (right) and unknown (grey).

WHY IT **MATTERS**

The periodic table sits both symbolically and practically at the heart of chemistry. It is no coincidence that the TV show *Breaking Bad* made use of the periodic table symbols for bromine and barium in its notable titles: the appearance of the distinctive "tiles" from the table with associated atomic number and weight immediately brings the viewer into the world of chemistry. There surely is not a school chemistry lab around the world without a periodic table chart on the wall.

"*BREAKING BAD* MADE USE OF THE PERIODIC TABLE SYMBOLS FOR BROMINE AND BARIUM IN ITS NOTABLE TITLES"

Practically speaking, the table is rich in information. At the basic level, each tile on the table contains the one- or two-letter chemical symbol, atomic number and weight for each element. Having a quick reference can be valuable, because the symbols themselves are not always obvious. Some make use of Latin or Greek names (for example, lead's Pb for *plumbum* in Latin and mercury's Hg for *hydragyrum* in Greek). Others can catch an English speaker out where, for example, tungsten is represented by W, after the element's German name *wolfram*. More recent elements may simply be unfamiliar, such as Ts for tennessine.

However, the table also shows significantly more, both in the relationship between elements in groups, and in the way various types of element clump together: different types of metal, the halfway house "metalloids" and non-metals. Just as the standard model of particle physics gives a core picture for much of physics, the periodic table is a starting point for understanding much of chemistry.

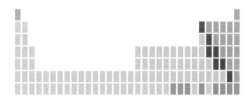

FUTURE **DEVELOPMENTS**

Over the years, new synthetic elements have been added to the table. Although the current heaviest element is oganesson, in principle others could be synthesized, though the elements are increasingly unstable. This is because as elements become heavier, the nucleus becomes larger. The strong nuclear force holding the nucleus together is extremely short range, and the size of nucleus for these heavier elements is far too big to be stable.

Apart from extending the table, there have been a number of attempts to find a better way to present the information, as the current structure with the lanthanides and actinides outside of the main layout is clumsy, and there is no specific reason why the table has to be presented in its current form. Alternative layouts may simply incorporate these rows, restructured to reflect the atomic structure more directly, or go for something more dramatic, such as three-dimensional or spiral-structured forms. For the moment, though, the classic table structure remains the standard.

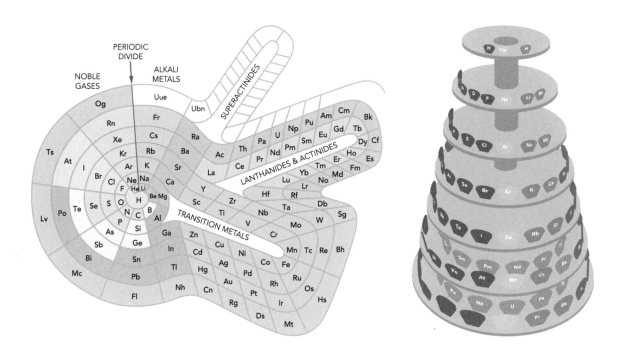

Above left: Alternative spiral periodic table structure developed by Otto Benfey.
Above right: A three-dimensional periodic structure.

THE **ESSENTIAL** SUMMARY

ORIGINS	KEY THEORIES AND EVIDENCE	CRITICS	WHY IT MATTERS	FUTURE DEVELOPMENTS
1803 John Dalton lists some elements in order of increasing atomic weight. **1829** Johann Döbereiner puts elements in groups of three with similar properties. **1864** John Newlands opts for octaves, returning to elements with similar properties with an expanded list. **1869** Dmitry Mendeleev publishes his first periodic table. Within two years he predicts the existence of some missing elements.	The **number of electrons in the outermost shells** in an element determine its chemical properties and position in the table. The clumsy-looking structure of the table reflects the way that **different electron shells can hold different numbers of electrons** – and each starts another shell when reaching 8 outer electrons even if the shell isn't full. None of the **elements above uranium (92)** naturally occur on Earth and none above plutonium (94) in space. Initially the table was built on **atomic weight, chemical properties** and **valence**. The table is **not always exactly sequential** on atomic weight. Modern information is derived from **electron shell structure** and weights from the **energy required to accelerate** a charged particle.	Early attempts were correctly criticized for the **arbitrary nature of assuming an octave**, as Newton did with rainbow colours. Mendeleev's approach was driven by logical observations and **made predictions** that could be tested. Despite its significance, Mendeleev **failed to win the 1906 Nobel Prize**, which was awarded instead for an inferior discovery as a result of political machinations.	The periodic table is **symbolically and practically at the heart** of chemistry. The table is rich in information, providing **chemical symbol, atomic number** and **atomic weight**. The relationship between **elements in groups** and the way types of element such as **non-metals** cluster together is highly instructive.	**New synthetic elements** could still be added, though they are likely to be **very unstable** as their nuclei are too large. **Alternative ways to present the information** are always being tried, as the traditional structure is rather clumsy. As yet nothing better has been found, but **different structures, including 3D**, could make the table more approachable.

BONDING AND CHEMICAL REACTIONS

THE **ESSENTIAL** IDEA

"THE CHEMICAL NATURE OF A COMPLEX MOLECULE IS DETERMINED BY THE NATURE OF THE ELEMENTARY COMPONENT PARTS, THEIR QUANTITY AND *CHEMICAL STRUCTURE*."

ALEKSANDR BUTLEROV, 1861

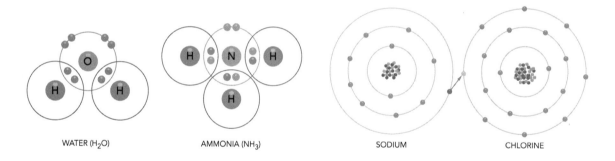

WATER (H$_2$O)   AMMONIA (NH$_3$)   SODIUM   CHLORINE

A major part of chemistry is concerned with the way that different atoms bond with each other to make compounds in processes known as chemical reactions. Chemical bonds are electromagnetic and come in two broad forms. Ionic bonds can form when the two reagents (chemicals that react with each other) are ions. So, for example, in sea water there are positively charge sodium ions and negatively charged chlorine ions, unconnected. When the water is evaporated, the positive sodium is attracted to the negative chlorine and the two are bonded together by the electrical charges to form sodium chloride – common salt.

The second kind of bond is covalent. Here the bond is formed by electrons that are shared between the atoms that form the bond. So, for example, natural

Above: Covalent bonds in the structures of water and ammonia, and the ionic bond in sodium chloride.
Opposite top: Jöns Jakob Berzelius (1779–1848).
Opposite bottom left: Gilbert Lewis (1875–1946).
Opposite bottom right: Title page of Newton's *Opticks.*

gas, methane, has four hydrogen atoms bonded to a carbon atom by covalent bond. The electrons that form the bond – one from each atom – are shared between the atoms that are bonded. So, each hydrogen atom shares a pair of electrons with the carbon atom.

"A MAJOR PART OF CHEMISTRY IS CONCERNED WITH THE WAY THAT DIFFERENT ATOMS BOND WITH EACH OTHER TO MAKE COMPOUNDS IN PROCESSES KNOWN AS CHEMICAL REACTIONS. "

ORIGINS

Alchemists had long known that elements combined together to make compounds. Isaac Newton, in his 1704 book *Opticks*, which contained a number of "queries" where he speculated on more general scientific matters, considered how atoms stuck together. There had been ideas dating back to the ancient Greeks that atomic bonding was due to special shapes of the atoms that allowed them to hook onto each other or some kind of gluing action, but Newton (correctly) put it down to an attractive force that was extremely strong over small distances, but did not extend far.

JOHN DALTON: THE FATHER OF ATOMIC THEORY

In 1803, John Dalton identified elements and a number of compounds, though some of his "elements" were in fact the names of compounds, while, as we have seen, his assumption that compounds would be formed by the simplest possible combination – making water, for example, HO rather than H_2O – was incorrect. Soon after, by 1819, the Swedish chemist Jöns Jakob Berzelius had made the first suggestion that the bonds between atoms were electrical in nature.

During the nineteenth century, noting how different atoms combine in different ratios, the concept of valence was combined with the idea of attracting electrical charges. The two main types of bonding would be first specifically identified in 1916. American chemist Gilbert Lewis devised the electron pair, or covalent, bond, while German physicist Walther Kossel came up with the theory of the ionic bond.

With the development of quantum theory and a better idea of atomic structure, by the 1930s the relationship of the distribution of outer electrons to bonding would be understood along with the quantum physics behind the structures of compounds.

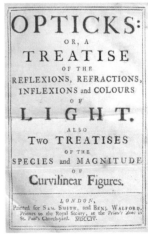

OPTICKS:
OR, A
TREATISE
OF THE
REFLEXIONS, REFRACTIONS,
INFLEXIONS and COLOURS
OF
LIGHT.
ALSO
Two TREATISES
OF THE
SPECIES and MAGNITUDE
OF
Curvilinear Figures.

LONDON,
Printed for Sam. Smith, and Benj. Walford,
Printers to the Royal Society, at the *Prince's Arms* in
St. Paul's Church-yard. MDCCIV.

KEY **THEORIES** AND **EVIDENCE**

IONIC, COVALENT AND FULL SHELLS

"REAGENTS ARE REGARDED AS ACTING BY VIRTUE OF A CONSTITUTIONAL AFFINITY EITHER FOR ELECTRONS OR NUCLEI... THE TERMS ELECTROPHILIC (ELECTRON-SEEKING) AND NUCLEOPHILIC (NUCLEUS-SEEKING) ARE SUGGESTED."

CHRISTOPHER INGOLD, 1933

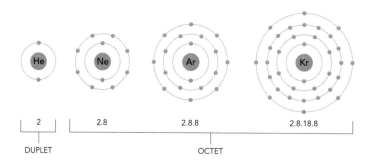

The early chemists and their predecessors the alchemists realized that not every possible combination of elements occurred in practice. As the relative weights of atoms and compounds were measured it became possible to see how different atoms combined and to assign valence to the elements – their natural tendency to combine in particular ratios with other atoms. Originally, valence was based on the number of hydrogen atoms an element would connect with, but it became clear that what was happening was an atomic preference for outer shells containing eight electrons (with the exception of hydrogen and helium where the number was two).

TYPES OF BONDS

This explained why the noble gases, formerly known as the inert gases, were very difficult to react with anything. Each of them (helium, neon, argon, krypton, xenon and radon) has a full outer shell, so doesn't want to add or take away any electrons.

In an ionic bond, one element loses outer electrons until it has emptied its outer shell, leaving it with the next shell full. This element becomes positively charged. The second element gains outer shell electrons to achieve a full set, making it negatively charged. In the simple example of sodium chloride, sodium loses one outer electron, leaving it with the same electronic structure as neon, while chlorine gains an outer electron, giving it the electronic structure of argon. The two electrically charged ions are stable as a result, and can be attracted together by their charges to make them stick together, just as an electrically charged balloon can pick up pieces of paper.

Above: The most stable forms have eight outer shell electrons (apart from helium).
Opposite top left: The methane molecule.
Opposite top right: Electrical attraction with a charged balloon.
Opposite bottom: Hydrogen bonding between water molecules.

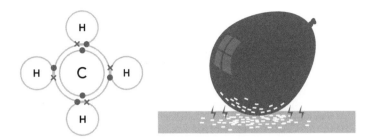

A covalent bond enables electrons to be shared between the bonded elements. So, for example, carbon has four electrons in its outer shell. It is this half-full aspect that makes carbon so flexible in its bonding, making the many different organic molecules, from the simplicity of methane to huge molecules such as DNA. Hydrogen, by comparison, has one electron in its outer shell.

When four hydrogen atoms bond with a carbon atom, all the hydrogens' outer electrons are shared by the carbon, adding four to its existing four, taking carbon up to the stable electronic structure of neon. Each hydrogen atom shares one of the carbon's outer electrons, taking the hydrogen atoms to the stable electronic structure of helium. Because electrons don't have a specific location but exist in a cloud of probability, the electrons can truly be said to be in both the carbon and the hydrogens' shells.

Bonds are not limited to single connections. Ionic bonds can have multiple extra or removed outer electrons in the outer shell, while there can be multiple covalent bonds between the same two atoms, balancing up the electron numbers in the outer shells.

WEAK BONDS

There are a number of other bonds. Metals have a variant of covalent bonding where free electrons can move through the lattice, conducting electricity and heat. There are also a number of partial or weak bonds, the best-known example being hydrogen bonds. These reflect the way that some molecules have a relative charge on particular atoms within the molecule. So, for example, water has a relative negative charge on its oxygen atom and relative positive charges on the hydrogen atoms.

Hydrogen bonding results in molecules being attracted to each other – not as firmly as a strong bond, but enough to have a significant impact. If it weren't for the hydrogen bonding preventing water molecules from easily flying away from each other, water would boil at –94°F (–70°C), meaning there would be no liquid water and hence no life on Earth. It is also why water is such a good solvent.

The evidence behind our understanding with bonds is very much the same as that for the atomic structure, as the two are so strongly tied together.

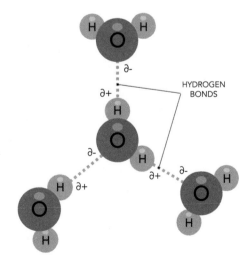

HYDROGEN BONDS

CRITICS

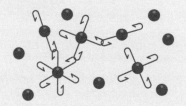

Without an understanding of bonding or a modern concept of atoms, the alchemists and early chemists thought that some substances had an "affinity" for others, meaning that they had a tendency to combine and react. Like much alchemy, the theory here was as much spiritual as it was scientific. As this was before the nineteenth-century introduction of the modern concept of an atom, the consideration was only of a quantity of substance. Similarly, these substances were not necessarily thought of as the elements we now know: the familiar elements were gradually introduced as they were isolated from compounds.

> "LIKE MUCH ALCHEMY, THE THEORY HERE WAS
> AS MUCH SPIRITUAL AS IT WAS SCIENTIFIC."

While not exactly criticism, there was considerable confusion in the nineteenth century as atomic theory gradually became accepted about the difference between atoms and molecules. While Dalton, for example, recognized that compounds were molecules (sticking to a "hooked together" model), he did not accept that elements themselves could come in compound form – so, for example, Dalton envisaged that the oxygen in the atmosphere was made up of individual atoms, where we now know it is in the form of oxygen molecules O_2.

Above top: An early concept of molecules formed when hooked atoms link.
Above bottom: Early chemical affinity table.
Opposite: The molecular structure of DNA.

WHY IT **MATTERS**

"IF THE PERIODIC TABLE IS THE HEART OF CHEMISTRY, CHEMICAL BONDS AND REACTIONS ARE ITS FOOD AND DRINK.."

Chemical reactions are central to everyday life. Until recently, most of our sources of heat and light were the result of combustion – a chemical reaction – and it is also a form of (mild) combustion that occurs in the digestive system to extract energy from our food.

Whether it is in a truly simple chemical reaction like an ionic bond forming as sodium chloride is precipitated from salt water, or in the intensely complex reactions of organic chemistry that lie behind the mechanisms of life – for example, those that make use of the structure of DNA – it is chemical bonds and the reactions involved in forming and breaking those bonds that are underway.

If the periodic table is the heart of chemistry, chemical bonds and reactions are its food and drink. Everywhere from a chemistry class at school to a world-class chemical laboratory – or the chemical reactions of nature – the making and breaking of bonds is essential for practically all activities. Chemical reactions are critical for many aspects of industry and every natural process.

As we have seen, the hydrogen bond in water is the reason that liquid water exists on Earth – and the chemical bonds in molecules are responsible for every structure that goes beyond individual atoms. These are amongst the key building blocks of reality.

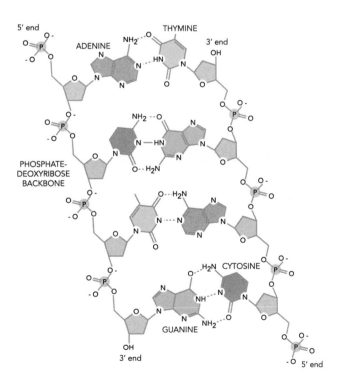

FUTURE **DEVELOPMENTS**

Chemical reactions and bonds are now extremely well understood; however, chemists are always pushing the boundaries of what is possible, extending the natural tendency of elements to bond together in unusual ways.

A good example is in experiments that try to make silicon parallel carbon. All known life is based on carbon, because its four-way bonding makes carbon the most flexible of elements for making a whole range of structures. However, the periodic table tells us that other elements in the same group have the potential for similar reactions as they too have four electrons in their outer shell. The closest possibility is silicon, and there has been speculation in the past that there could be silicon-based lifeforms, with silicon taking on the role of carbon.

In practice, though, silicon isn't up to the job. In 2009, researchers at Imperial College, London managed to make a short-lived equivalent of a classic organic carbon structure, the benzene ring, which has six carbon atoms in a hexagonal ring with attached hydrogen. The silicon version was distorted and unstable. According to the researchers, what's stable for carbon isn't for silicon. It is only with this kind of ongoing research into different new materials that we can discover new possibilities for chemical reactions and bonds.

Above top: The silicon equivalent of benzene above benzene's carbon structure.
Above bottom: Sea salt drying after forming ionic bonds between sodium and chlorine ions.

THE **ESSENTIAL** SUMMARY

ORIGINS	KEY THEORIES AND EVIDENCE	CRITICS	WHY IT MATTERS	FUTURE DEVELOPMENTS
5th century BC – Ancient Greek philosophers suggest atoms hook together to form substances.	The **number of electrons in the outermost shells** in an element determine how an element will bond.	Early chemists and alchemists thought that different substances had **an affinity** for each other, encouraging them to react.	Chemical reactions are **central to everyday life**.	**Reactions and bonds** are extremely well understood**.**
1704 – Isaac Newton suggests atoms are attracted to each other by a very short-range force.	Elements bond in such a way to get to **eight electrons** in the outer shell.	Initially there was considerable confusion between atoms and molecules. Dalton, for example, did not realize **a single element could exist a molecule** rather than individual atoms.	Breaking and making chemical bonds is involved in everything **from salt forming to DNA mechanisms**.	Experiments continue to be made into **different possible molecular structures**. For example, experimenting with **using silicon in place of the carbon** in organic molecules. It turns out not to work – but it had been speculated that silicon could also support life.
1803 – John Dalton identifies elements and some compounds.	In an **ionic bond** one atom loses one or more electrons and another gains electrons to achieve a full outer shell and the two are electrically attracted.		**Every natural process** (other than nuclear reactions) involves chemical bonds.	
1819 – Jöns Jakob Berzelius suggests chemical bonds are electrical.	In a **covalent bond** a pair of electrons are shared between two atoms to fill up their outer shells.		**Every structure** beyond an individual atom involves chemical bonds.	
1916 – Covalent and ionic bonds are identified.	Ionic bonds can have **multiple electrons added or removed**, while two elements can be joined by **more than one covalent bond**.			
1930s - quantum theory and atomic structure developed sufficiently to understand the role of the outer shell of electrons.	Other bonds include the **metallic bond** and the **hydrogen bond**.			

ORGANIC AND INORGANIC CHEMISTRY

THE **ESSENTIAL** IDEA

"I AM AN ORGANIC CHEMIST, ALBEIT ONE WHO ADHERES TO THE DEFINITION OF ORGANIC CHEMISTRY GIVEN BY THE GREAT SWEDISH CHEMIST BERZELIUS, NAMELY, THE CHEMISTRY OF SUBSTANCES FOUND IN LIVING MATTER, AND MY SCIENCE IS ONE OF THE MORE ABSTRUSE INSOFAR AS IT RESTS ON CONCEPTS AND EMPLOYS A JARGON NEITHER OF WHICH IS PART OF EVERYDAY EXPERIENCE."

ALEXANDER TODD, 1957

There is a great divide in chemistry, between organic and inorganic. As "organic" has been taken over as a marketing label for some types of agricultural produce, it can be a confusing term. An early definition by the German chemist Friedrich Kekulé von Stradonitz was simply that organic chemistry was the chemistry of carbon compounds.

Some would now argue whether all carbon compounds, such as the greenhouse gas carbon dioxide, are organic – but certainly organic compounds all contain carbon. By Swedish chemist Jöns Jakob Berzelius's definition, we might expect something like, say, a calcium compound in animal bones to be considered organic. Probably the best definition combines the two and makes organic chemistry that of carbon compounds found in living matter.

By contrast, inorganic chemistry – which was most of chemistry until recent times – is easy to define. It is any chemistry that does not involve organic substances.

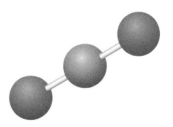

Right: Carbon dioxide – contains carbon but not usually considered organic.
Opposite top: Robert Boyle (1627–1691).
Opposite bottom: The four elements of Empedocles.

ORIGINS

The fifth-century BC ancient Greek philosopher Empedocles devised four elements – earth, air, fire and water – which he believed made up everything. This had some logic – for example, when a log burned, the earthy wood oozed a liquid, gave off airy smoke and fire. Aristotle, in the fourth century BC, built on this idea, adding a fifth element for the heavens. Although these elements would still be regarded as real for around 2,000 years, it became increasingly clear through the works of alchemists that on a practical level there was a wider range of substances that acted as fundamental elements.

As the word suggests, chemistry grew out of alchemy. Many aspects of alchemy were unscientific, based on assumptions of the ability to find a "philosophers' stone" that turned lead to gold, or the elixir of life – this was the so-called "operative" alchemy, which was, for example, a lifetime study of Isaac Newton. Although we remember Newton for his physics, he had more books in his library on alchemy than either maths or physics. This aspect of alchemy would fade away as understanding of substances grew.

However, the other side of alchemy, "speculative" alchemy, was more concerned with how the elements (increasingly recognized as involving far more than the four ancient Greek elements) combined to produce different substances. A leading proponent of speculative alchemy who ushered in the new science of chemistry was Robert Boyle, whose book *The Sceptical Chymist* could be said to have begun the field.

The distinction between organic and inorganic chemistry would not come until the nineteenth century. The first recorded use of the word "inorganic" seems to have been in 1794, when an R. J. Sullivan wrote, "We may safely conclude that the mineral kingdom, that assemblage of brute inorganic bodies… has yet distinct families and species." Here "inorganic" meant literally lacking organs, which seemed a marker of living things.

By 1831 the term organic was evolving and an anatomy book could comment, "In living bodies there are two kinds of elements, inorganic and organic. The inorganic elements are those which may be obtained by the processes of chemistry from minerals as well as organized bodies." Two years later, Berzelius made the clear distinction between the carbon compounds of living matter and inorganic compounds.

AIR

FIRE

WATER

EARTH

KEY **THEORIES** AND **EVIDENCE**

ATOMS, COMPOUNDS AND ORGANICS

Chemistry is primarily the study of elements and compounds, where compounds are substances made up of molecules featuring two or more different elements. The inorganic compounds are all those lacking carbon – the bulk of compounds on Earth, as the majority of the planet is inorganic.

Although there is some dispute over where to draw the line, a number of simple carbon compounds, such as carbon dioxide and calcium carbonate, are usually considered inorganic. Inorganic substances form the majority of ionic-bond compounds and many of them when solid have regular crystal structures.

Organic compounds contain carbon and are mostly associated with life, either in being present in living organisms or (for example in fossil fuel compounds) being produced from decaying lifeforms. The chemistry of organic compounds tends to be more complex than inorganic chemistry because of the flexibility of carbon. As we have seen, the way that elements react with each other is determined by the electrons in an element's outer shell. Carbon has four electrons and four vacancies, contributing to its unique effectiveness at building complex structures.

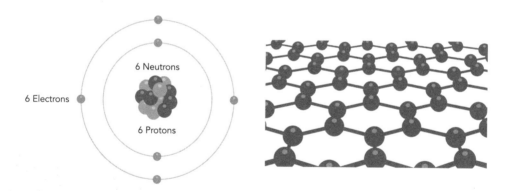

6 Electrons

6 Neutrons

6 Protons

Carbon atoms make very effective bonds, carbon-to-carbon. These bonds are present in natural forms of carbon, such as graphite and diamond. However, they come into their own in compounds. Rather like a construction set of identical parts, carbon atoms can be linked together in chains. If each carbon atom has a single bond to another and the rest of its vacancies are filled with hydrogen we get the alkane family – after the simplest, methane, comes ethane with two carbons, propane with three carbons, butane with four carbons and so on.

MULTIPLE BONDS

However, organic chemistry has more tricks up its sleeve than simply chaining carbon atoms together with single bonds. There can be double or even triple bonds between pairs of carbon atoms. The simplest double-bonded equivalent to alkanes (known as alkenes) has a pair of carbon atoms with a double bond between them and four hydrogen atoms attached.

With the ability to have either a single or double bond comes a speciality of organic chemistry, the aromatic compounds. These involve special ring structures that are particularly chemically stable. (Organic compounds that are not aromatic are called aliphatic.) The definitive such ring structure is benzene. Here, six carbon atoms join together in a hexagon, nominally featuring alternating single and double bonds

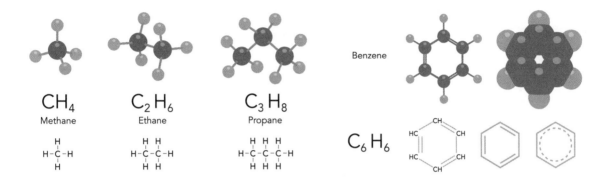

CH_4
Methane

$$H-\overset{\displaystyle H}{\underset{\displaystyle H}{C}}-H$$

C_2H_6
Ethane

$$H-\overset{\displaystyle H}{\underset{\displaystyle H}{C}}-\overset{\displaystyle H}{\underset{\displaystyle H}{C}}-H$$

C_3H_8
Propane

$$H-\overset{\displaystyle H}{\underset{\displaystyle H}{C}}-\overset{\displaystyle H}{\underset{\displaystyle H}{C}}-\overset{\displaystyle H}{\underset{\displaystyle H}{C}}-H$$

Benzene

C_6H_6

between the carbon atoms, leaving space for six hydrogen atoms to be attached. The reason that aromatic rings are very stable is down to the quantum nature of electrons. Rather than actually forming alternating single and double bonds, the electrons average out around the ring, effectively providing 1½ electrons per bond.

As well as forming rings, organic compounds are great at building long chains, known as polymers. Most plastics are made up of long chain polymers, and polymers also turn up in nature. They form the compounds that give rigidity to organic structures and a special kind of polymer provides the structure for the most important of all organic compounds: DNA.

We'll come back to DNA in the genetics section of biology, but in terms of its organic chemistry, DNA (deoxyribonucleic acid) is not a single compound, but a whole family. It consists of the familiar dual helix, a pair of polymers made of chains of sugar molecules, linked by a series of pairs of organic molecules which look like the treads in a spiral staircase. These are the "base pairs" where DNA's information is stored. Each chromosome, organic compounds found in most cells, is a single DNA molecule, which includes some of the largest molecules in existence. Our biggest chromosome, mid-sized in the natural world, contains around 10 billion atoms.

EVIDENCE

The structures of organic and inorganic molecules were initially deduced from a combination of the weights of elements present and a growing understanding of the structure of the atom and chemical bonds. Some molecular structures, such as DNA, would be discovered using X-ray diffraction, where X-rays are scattered by the atoms to produce patterns, while more information was provided by spectroscopy.

Opposite: Carbon's atomic structure and lattice structure in graphite.
Above left: Aliphatic organic compounds.
Above right: Benzene, the definitive aromatic organic compound.
Right: X-ray diffraction image from DNA.

CRITICS

"TO MANY CHEMISTS IN THE 1920S AND EARLY 1930S, THE ORGANIC CHEMIST WAS A GRUBBY ARTISAN, ENGAGED IN AN UNSYSTEMATIC SEARCH FOR NEW COMPOUNDS, A SEARCH WHICH WAS STRONGLY INFLUENCED BY THE PROFIT MOTIVE."

LOUIS HAMMETT, 1966

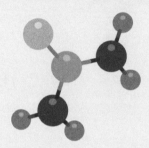

The early distinction between inorganic and organic chemistry was based on the idea that living things were powered by some special internal energy, the so-called "vital force", which imbued organic compounds with a special nature and drove organic chemistry. This idea, which was supported by Berzelius, has lingered on in some alternative medicines, such as traditional Chinese medicine, but in the chemical world has been dismissed.

It had been argued that because of the lack of vital force it would be impossible to make an organic compound from inorganic substances – yet this was first countered in the 1820s, when German chemist Friedrich Wöhler produced urea, an organic compound of carbon, oxygen, nitrogen and hydrogen, from inorganic components. Vitalism proved difficult to shake and dragged on into the early twentieth century, but by the 1920s it was on its last legs.

As the quote above suggests, some "pure" chemists originally saw organic chemistry as little more than a cookery exercise in constructing new compounds in the hope that some would be profitable. However, by the 1940s and 1950s, as the sophistication of organic structures was realized, it became clear that there was much to learn at the theoretical level too.

WHY IT **MATTERS**

The chemistry of the two branches, inorganic and organic, tends to be structurally separate, although, of course, an organic chemist may well make use of inorganic compounds. Although there are far fewer common inorganic compounds than organic, inorganic chemistry includes many familiar chemical substances, from common salt to water and sulfuric acid. The chemical reactions of these substances are often relatively simple and important in gaining a basic understanding of chemistry. And inorganic substances are the primary constituents of the manufactured world, from concrete and steel to the microscopic building blocks of electronics in silicon chips.

"ORGANIC CHEMISTRY IS THE CHEMISTRY OF LIFE, AND SO FEATURES STRONGLY IN BOTH MEDICINE AND OUR UNDERSTANDING OF BIOLOGY."

However, there is little doubt that the dominant side of the two branches is now organic. This is in part because of the scale of organic chemistry, both in the numbers of possible organic compounds and the sheer baroque complexity of some organic molecules. But also it is because organic chemistry is the chemistry of life, and so features strongly in both medicine and our understanding of biology.

Medicine has come to be dominated by molecular biology – an understanding of how biological organisms work on the level of the interaction of different molecules and, fundamentally, this is pure organic chemistry. Modern medicine is closely tied to organic chemistry. Although now regarded with mixed feelings, organic chemistry is also the chemistry of the petrochemical industry, from plastics to aviation fuel and petrol.

Opposite top: Urea molecule, synthesized by Wöhler.
Opposite bottom: The elements of living things were long thought different from inorganic substances.
Above left: Chemistry is vital for the built environment.
Above right: Organic chemistry dominates the biomedical sciences.

FUTURE **DEVELOPMENTS**

Some of the biggest developments in organic chemistry come from our ability to deal with individual molecules. One of the first powerful examples of this was in DNA fingerprinting, where DNA is cut into fragments using enzymes (proteins that act as catalysts) to enable small sections to be compared. But such work on DNA is relatively minor when compared with the latest ability to manipulate organic compounds using a technique known as CRISPR (clustered regularly interspaced short palindromic repeats), which employs a mechanism found in the immune system to target specific genes, using it to cut and splice DNA as if it were a piece of movie film.

The possibilities for using CRISPR are still being fully discovered. At the same time, we are gradually finding out more about the remarkable molecular machinery employed in biological organisms, providing molecule-sized equivalents of rotary motors, ratchets and other constructs. Organic chemistry still has much to be discovered.

Above: CRISPR enables scientists to cut and splice DNA.

THE **ESSENTIAL** SUMMARY

ORIGINS	KEY THEORIES AND EVIDENCE	CRITICS	WHY IT MATTERS	FUTURE DEVELOPMENTS
c444 BC Ancient Greek philosopher Empedocles devises the four elements: earth, air, fire and water.	Inorganic compounds include all those **not featuring carbon**, making up most of the Earth.	Originally it was assumed that organic chemistry required a **"vital force"** that gave an organism life.	Inorganic compounds make up many **familiar substances**, including common salt, water and sulfuric acid.	Our ability to **manipulate individual organic molecules** is becoming increasingly effective.
350 BC Aristotle extends Empedocles's theory, adding a fifth element for the heavens.	A few simple carbon compounds, such as **carbon dioxide and carbonates**, are considered inorganic.	The first challenge to the vital force was when an **organic compound**, urea, was **synthesized from inorganic substances**.	Inorganic chemistry is central to the **manufactured world** and the microscopic world of **electronics**.	DNA fingerprinting involves the use of **enzymes to chop up DNA** into fragments for comparison.
1661 Robert Boyle publishes *The Sceptical Chymist,* marking the separation of chemistry from alchemy.	Carbon's flexible structure with **four electrons and four vacancies** in its outer shell make it uniquely capable of forming large, complex molecules.	**Vitalism** would linger on to the 1920s.	Organic chemistry is the **chemistry of life**.	We are just starting to understand the powerful capabilities of **CRISPR**, a technology used to cut and splice DNA.
1794 R. J. Sullivan makes use of the term "inorganic".	Carbon–carbon bonds can be **single, double or even triple**. Alternating single and double bonds in a hexagonal ring makes **benzene, the simplest aromatic compound**.	Otherwise, the main controversy remains over exactly **where to draw the line** between organic and inorganic.	Medicine is now dominated by **molecular biology** – understanding the chemistry of organic molecules.	There is still much to learn about the complex **organic molecular machinery** used in living cells.
1833 Jöns Jakob Berzelius distinguishes organic and inorganic chemistry.	Organic compounds can make **long-chain polymers**, essential for plastics and forming the dual-helix backbone of DNA.		Despite being regarded with mixed feelings now, organic chemistry is the **chemistry of petrochemicals**, from plastics to petrol.	
	Molecular structures are often discovered using **spectroscopy and diffraction** techniques.			

INDUSTRIAL CHEMISTRY

THE **ESSENTIAL** IDEA

"LIEBIG TAUGHT THE WORLD TWO GREAT LESSONS. THE FIRST WAS THAT IN ORDER TO TEACH CHEMISTRY IT WAS NECESSARY THAT STUDENTS SHOULD BE TAKEN INTO A LABORATORY. THE SECOND LESSON WAS THAT HE WHO IS TO APPLY SCIENTIFIC THOUGHT AND METHOD TO INDUSTRIAL PROBLEMS MUST HAVE A THOROUGH KNOWLEDGE OF THE SCIENCES."

IRA REMSEN, 1910, REFERRING TO JUSTUS VON LIEBIG

Chemistry typically brings one location to mind: a chemistry lab, with test tubes, jars of chemicals and Bunsen burners. It's what chemistry was like at school. However, the vast majority of the world's chemistry takes place not in a laboratory but on an industrial site.

Here, the reaction vessels are more likely to be huge stainless steel cylinders than glass tubes, and the quantities involved can be astonishing – but the same chemical principles are at play as shape the chemistry of the laboratory bench.

Industrial chemistry is involved at some stage in the production of manufactured goods, pharmaceuticals, foods and more. It's unlikely that there is a room in your house that does not contain the output of industrial chemistry, from the dyes that colour fabrics to the plastics, glass, metals and ceramics that make up products.

ORIGINS

Although some suggest that the production of pigments for prehistoric wall art was early industrial chemistry, the earliest known true practical chemistry was the industrial application in the production of metals from ores. While native copper and meteoric iron were used in prehistoric times, copper and tin-bearing ores for bronze, and lead and iron ores were being converted into metal well into prehistory – in some cases as much as 8,000 years ago.

Exactly how the smelting process was discovered is not clear, as it predates written records. Generally, an ore, which is typically a mix or a complex compound, would be heated in air to produce a simpler compound, often the oxide of the metal. This would then have to be raised to a higher temperature with a substance that reacts with oxygen (a reduction reaction) to produce the pure metal.

NEW WINE

Other early industrial processes included the production of fermented products such as beer and wine. Unlike a modern chemist, these early chemical workers would not be aware of why they achieved the output that they did. A chemical process is likely to have originally been stumbled across by accident, but once the results had been observed it would be repeated and refined until it produced the desired outcome.

It wasn't until the sixteenth century that there was any attempt at what might be considered a scientific take on industrial chemistry – alchemy was typically carried out more on an individual level. But in 1556, a book called *De re metallica* (*On the Nature of Metals*) by the German metallurgist Georg Pawer, better known by his Latinized name Georgius Agricola, became the definitive tome on metal mining and extraction.

It would be nearly 200 years later, with the birth of the industrial revolution, that industrial chemistry moved beyond its ancient remit and expanded greatly, with the production of valuable chemical compounds such as sulfuric acid and lime (calcium hydroxide) being undertaken in purpose-built factories, rather than home workshops. One of the first such factories producing sulfuric acid was opened in Prestonpans in Scotland in 1749. Industrial chemistry would accelerate in significance with the development of new synthetic compounds, such as mauveine, the first successful synthetic dye, produced by William Perkin in 1856 by accident while trying to produce artificial quinine.

Opposite: Industrial plants undertake chemistry on a massive scale.
Top: Georg Pawer (1494–1555).
Bottom: Early French chemical plant vulcanizing rubber.

KEY **THEORIES** AND **EVIDENCE**

REDOX, HABER–BOSCH AND CATALYSIS

"WHEREVER WE LOOK, THE WORK OF THE CHEMIST HAS RAISED THE LEVEL OF OUR CIVILIZATION AND HAS INCREASED THE PRODUCTIVE CAPACITY OF THE NATION."

CALVIN COOLIDGE, 1924

The theories behind the reactions of industrial chemistry are exactly the same as those behind laboratory-based chemistry: those that we met in the bonding and chemical reactions section. Because industrial chemistry is so broad in scope, it is difficult to cover all the key reactions without resorting to a tedious list. Rather than do this, it would be best to give a little detail behind a handful of the most common or most important reactions and processes used in industrial chemistry.

Many industrial chemical processes involve a redox reaction, even if they aren't purely dependent on such a reaction. The term "redox" is a contraction of "reduction and oxidation", which occurs when one of the chemicals involved (reagents) is reduced and one is oxidized. Originally, as their name suggests, oxidation meant reacting with oxygen, resulting in an oxide forming, but now the term is generalized to a reaction where a substance (the "oxidizing agent") accepts electrons from the substance that is oxidized. Reduction is what occurs when a substance gains electrons. Originally this meant that a substance lost oxygen and so reduced in weight – it was reduced from, say, a metal oxide to a metal. In many reduction reactions (particularly organic reactions), hydrogen is the reducing agent of choice.

THE HABER-BOSCH PROCESS

If we were to pick out one chemical industrial reaction that in its time has had the biggest impact, one contender would have to be the Haber–Bosch process. It was responsible for both the ability to feed a growing world population through the development of fertilizers and the ability to create explosives. In both cases, the central chemical element was nitrogen.

Nitrogen is available in abundance – it makes up around 78 per cent of the Earth's atmosphere. Although plants are great at getting carbon out of the atmosphere for the essential molecules of life, they are less effective at "fixing" nitrogen – converting it to the simple nitrogen compound ammonia (one nitrogen atom with three attached hydrogen atoms) and derivates such as nitrates, which can then be used in constructing the essential biological building blocks known as amino acids. Plants are unable to do this, but bacteria that cluster round some plant roots, notably legumes and clovers, can do so. For agriculture to succeed on any scale, this natural process needs to be supplemented with nitrogen-rich fertilizer.

The Haber–Bosch process involves passing nitrogen and hydrogen at high temperatures and pressures over catalysts, typically fine particles of iron, which encourage the reaction. The development of the process to industrial levels was sparked by the First World War. Germany was cut off from the nitrates that were essential for production of explosives for munitions, and resorted to taking atmospheric nitrogen and converting it first to ammonia, which is then reacted in a multistage process to produce the nitrates that can act both as oxidizers in explosives and as fertilizer.

CATALYSTS

Catalysts also play a role in one of the most important reactions in the petrochemical industry: catalytic cracking. Crude oil first goes through another key industrial chemical process, fractional distillation. This is a way of separating out a liquid that is a mix of different volatile chemicals, which involves feeding the crude oil at a high temperature into the bottom of one of the tall metal cylinders often seen in industrial chemical sites. The lighter molecules rise up the cylinder, cooling, while the heavier molecules stay near the bottom. Feeds are taken out of the cylinder at different levels, resulting in splitting the oil into "fractions".

The output of the bottom of the column is then heated further and vaporized, and sprayed over extremely hot catalysts where the large, heavy molecules are split (cracked) into different petrochemical molecules such as butane, propane, petrol and fuel oil. These are separated by another fractional distillation.

Opposite left: Rusting is a classic redox reaction.
Opposite right: Fractional distillation cylinders at a refinery.
Top: Fritz Haber (1868–1934).
Bottom: In a munitions factory.

CRITICS

From its earliest days, industrial chemistry has involved handling hazardous materials. For example, back in the bronze age, before tin was used, bronze was made with arsenic, producing dangerous fumes in the smelting process. Not surprisingly, the industry has a mixed reputation – no one would enthusiastically live next to an industrial chemical plant – not helped by a history of littering the environment with waste products.

Perhaps the biggest example of public hazard arising from industrial chemistry was the Bhopal disaster in 1984. This took place at a Union Carbide pesticide factory in Bhopal in the Madhya Pradesh region of India. The plant stored large quantities of methyl isocyanate, a poisonous substance used in the production of pesticides. Water got into a tank containing around 40 tons of methyl isocyanate, resulting in a chemical reaction that produced sufficient heat to rapidly raise the temperature in the tank. The safety systems failed, resulting in the release of a large quantity of methyl isocyanate gas into the atmosphere. Over half a million people were exposed to the leak: of these over 150,000 were hospitalized and thousands died.

"OVER HALF A MILLION PEOPLE WERE EXPOSED TO THE LEAK: OF THESE OVER 150,000 WERE HOSPITALIZED AND THOUSANDS DIED."

While in some countries there is strong legislation to provide protection for workers and the environment, this is not universal. By far the biggest – and legitimate – criticism of industrial chemistry is where appropriate safeguards are not put in place. Governments and businesses are criticized rightly by environmental groups, though arguably the main burden lies with the governments that are not ensuring the safety of their citizens.

Above left: Environmental protest.
Above right: Ruins of the Bhopal plant.
Opposite: Piping in a chemical factory.

WHY IT **MATTERS**

Industrial chemistry lies at the heart of manufacturing, whether it involves producing the raw materials that will become manufactured goods or developing the pharmaceuticals and insecticides that help protect us from disease.

Look around you as you read this book. Perhaps you're at home – you'll probably see textiles made with artificial fibres; paints, varnishes and dyes from the chemical industry; objects incorporating metals and plastics or processed using chemicals. Even the book itself contains a range of components that are the result of chemical processes: the paper will have been treated chemically, the ink and the glue are both chemical products.

> "UNLESS YOU ONLY BUY ORGANIC, YOUR FOOD WILL HAVE BENEFITED FROM WEEDKILLER, PESTICIDES AND FUNGICIDES FROM THE CHEMICAL INDUSTRY"

In the kitchen, unless you only buy organic, your food will have benefited from weedkiller, pesticides and fungicides from the chemical industry (in reality, organic food indirectly benefits, as materials fed into the organic system originate this way and the organic industry would not be sustainable without this). Even the mortar that holds the bricks together or the grout in your bathroom are chemical products.

Alternatively, you might be on a train, bus or plane. Again, each will have benefited – both in its construction and in the materials used to keep it running – from the work of industrial chemists. There is hardly an aspect of everyday life that isn't improved by the output of the chemical industry.

FUTURE **DEVELOPMENTS**

Industrial chemistry will continue to play a major part in the world economy, but we can expect changes in emphasis. At the moment, petrochemical processes, both for fuel and plastic production, are important industrial chemical applications, but the move away from fossil fuels will see a significant reduction in demand. Although there has been a reaction against plastic as well, due to plastic pollution in the oceans, this is a less significant driver than the climate change that is behind the downswing in fossil fuel use – plastics are here to stay in many essential applications. Petrochemicals will also continue to be the starting point for a range of medical products.

Many of the chemicals used as building blocks in producing a wide range of products will continue as they have previously. Probably the biggest changes are likely to come in the pharmaceutical sector, where there is both a constant turnover of new products and we are seeing more potential for tailoring of pharmaceuticals to individual genetic requirements. Despite their high cost, pharmaceuticals are relatively small-scale in comparison with other industrial chemistry, and if personalized medication becomes commonplace we are likely to see more bespoke manufacturing, perhaps using forms of 3D printing, becoming part of the industrial chemistry landscape.

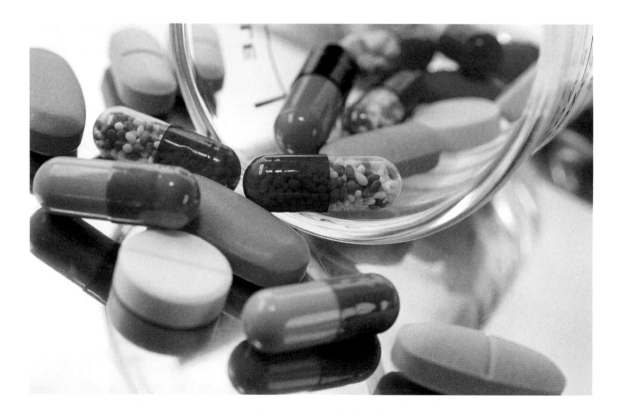

Above: Pharmaceutical production is changing faster than other sectors.

THE **ESSENTIAL** SUMMARY

ORIGINS	KEY THEORIES AND EVIDENCE	CRITICS	WHY IT MATTERS	FUTURE DEVELOPMENTS
c6000 BC Metal ores are first converted into metals using a smelting process. **1556** Georgius Agricola writes *De re metallica* on metal mining and extraction. **1749** One of the first chemical factories, creating sulfuric acid, was opened. **1856** William Perkin develops the first successful aniline synthetic dye, mauveine.	The theories here are those covered under **Bonding and Chemical Reactions**. Many industrial chemical processes incorporate **redox reactions** where electrons are moved from one substance to another. The **Haber–Bosch process**, converting atmospheric nitrogen to ammonia, transformed agriculture through the availability of cheap nitrate fertilizers (and helped with the production of explosives). **Fractional distillation and catalytic cracking** are central to extracting usable compounds from mixes such as crude oil.	Many industrial chemical processes are **potentially hazardous**: one of the first, producing bronze, involved arsenic, giving off dangerous fumes. Arguably the largest negative impact from the chemical industry was the **Bhopal disaster** in India, killing thousands and injuring many more. In some countries there is strong **protective legislation** for workers and the environment, but this is not universal. Governments and businesses are criticized by environmental groups for failing to provide appropriate legislation and checks, resulting in **pollution and health risks**.	Industrial chemistry is at the **heart of manufacturing**. In the home we are surrounded by **products that have involved industrial chemistry** in their production – even this book is made up of **paper and ink**, which have industrial chemistry behind them. Almost every aspect of **everyday life** has been improved by the products of the chemical industry.	With the move away from fossil fuels we will see **less use of crude oil**, though petroleum products will remain important in medical and plastic uses. Commodity chemicals will continue to be widely used as the **building blocks** for a wide range of products. In the **pharmaceutical sector** we will see continued development of new products as we get a better understanding of molecular biology, and more **individually tailored chemicals** for personalized medication.

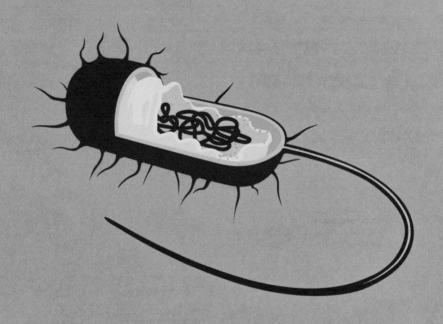

03

BIOLOGY AND
EVOLUTION

WHAT IS LIFE?

THE **ESSENTIAL** IDEA

"THE BROADEST AND MOST COMPLETE DEFINITION OF LIFE WILL BE – *THE CONTINUOUS ADJUSTMENT OF INTERNAL RELATIONS TO EXTERNAL RELATIONS.*"

HERBERT SPENCER, 1864

Biology is the study of living things – but, embarrassingly, it is surprisingly difficult to define just what life is. It is relatively easy to say that some things are alive (animals and plants, for example), while others are not (rocks and books, for example). However, there are plenty of examples that sit on the margins where it is not clear what the answer is.

Many definitions of life, like the one from Spencer above, are sufficiently vague and hand-waving that they prove almost impossible to use. Spencer's definition, for example, could arguably be used to suggest that a central heating system with a thermostat is alive.

Instead, science is often limited to identifying life through its properties and characteristics, listing, for example, the typical capabilities and actions of a living thing. Similarly, it has been suggested that the ability of life to keep a system in a non-equilibrium state is its defining quality. However we look at it, life is both fascinating and hard to pin down.

ORIGINS

Historically, life was often assumed to be the result of some special energy, soul or spirit. In some cultures, this energy was associated with air, which is why there are so many words linking air and spirit (think of respiration and expire, for example). This assumption is known as "vitalism" and was common both in Ancient Egypt and Ancient Greece. The seventeenth-century French philosopher René Descartes crystallized earlier ideas of separate mind and matter, known as dualism, which, though not identical, can also be seen as implying a separate and special life force.

By the twentieth century, vitalism had been scientifically confined to alternative medicine, leaving the idea that life was an emergent property of the physical components of a living organism with nothing extra required. Emergent properties are when the capabilities of a collection of components working together are greater than simply the sum of the parts, and are common in a range of natural systems.

A LIVING OBJECT

It is difficult to say exactly when the modern idea of a living object in the form of an organism came into use. The word "organism" has been used since at least 1701, but initially it meant simply having an organic nature – so, for example, the earliest recorded use refers to "the advantageous Organism of the eye", not a standalone living creature. In its modern sense, organisms have been referred to since the 1830s, which probably marks the change in attitude to a modern viewpoint, primarily driven by the behaviours of a living entity.

Two specific discoveries have proved particularly difficult for the identification of life: cells and viruses. Cells were discovered with the early use of microscopes in the seventeenth century. While singled-celled organisms were clearly seen to be alive, it would prove harder to say for certain if an individual cell from a multi-celled organism could be considered to be a living thing in its own right. Similarly, viruses, discovered around the beginning of the twentieth century (mostly being significantly smaller than bacteria), have some but not all the properties of living things and so remain on the margins of life.

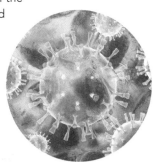

Opposite: The distinctive nature of life is
easily recognized but hard to define.
Top: René Descartes (1596–1650).
Bottom: Viruses have some but not all the properties of life.

KEY **THEORIES** AND **EVIDENCE**

PROPERTIES, STRUCTURES AND ENTROPY

Biologists resort to defining life by a number of properties of a living organism. There are often seven of these, though there are variants, so some of the headings below overlap:

- **Moving or responding to a stimulus** – we are used to animals moving, but plants move too, just on a different timescale, and all respond to physical and chemical stimuli.
- **Nutrition** – any living thing consumes energy, which has to come from somewhere. It may be by breaking down food to generate chemical energy, or creating chemical energy from sunlight through photosynthesis, but nutrition plays a part.
- **Respiration (sometimes combined with nutrition as metabolism)** – there are a number of processes that contribute to the production of energy, which often involve a reaction with oxygen and usually produce waste gases.
- **Organization** – all living things have some kind of internal structure, and all are based on the format of one or more cells, which can be prokaryotic (cells, such as bacteria, lacking a nucleus) or eukaryotic (more complex cells, such as in animals and plants, featuring a nucleus).
- **System regulation (homeostasis)** – as a living organism is a system that takes in and gives off energy in different forms, it is essential that it has feedback and control mechanisms to maintain internal properties.
- **Excretion** – the counter-property to nutrition. A living thing generates waste, both in terms of chemical content and heat, which needs to be removed from the system.
- **Reproduction** – a living organism is capable of producing copies of itself to maintain the species. Often this is not a direct copy, but involves some form of variation, increasing survival capability in changing environments.
- **Sensing** – a key survival requirement is being able to detect the surrounding environment, usually through sensors responding to different forms of energy.
- **Growth** – as they develop, organisms typically grow in size, complexity or both.

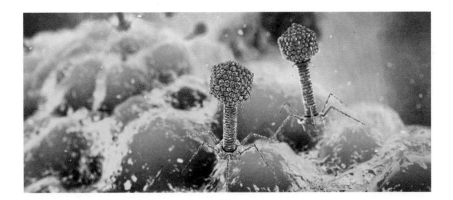

Arguably, many of these properties, such as having a cellular structure and system regulation, are typical rather than necessary – it would be possible to imagine a form of life that did not have them. Even reproduction is not necessary for a particular organism to be alive: there are plenty of organisms (mules, for example) that are unable to reproduce. By contrast, some structures have subsets of these properties but aren't alive. For example, a snowflake has structure, while a chemical reaction excretes heat. But there is a notable borderline case that has more but not all of the properties of life: a virus.

Where bacteria are definitely alive, viruses are significantly different. Usually much smaller than a bacterium, a virus lacks the mechanism for reproduction. Instead, it hijacks the cell-reproduction mechanism of an organism that it infects. For a long time, this meant that viruses were not considered to be alive. However, now the picture is more nuanced. In recognizing that the various properties are rules of thumb, some scientists accept that a virus is a specialist life form.

WHAT IS LIFE?

Although it seems outside their expertise, a number of physicists have taken an interest in the nature of life, since quantum physicist Erwin Schrödinger published a book called *What is Life?* in 1944. This looked at life from the point of view of energy flows and entropy.

As we have seen, entropy is an important physical quantity in thermodynamics, which can be seen as a measure of the disorder in a system. Living organisms are highly organized compared to a collection of atoms, implying a need to lower entropy to bring life into being. The natural tendency is for entropy to increase, but by sending heat out into its surroundings (known as a "dissipative process") a system can reduce entropy – this is effectively how a refrigerator works.

By transferring entropy from itself to its surroundings an organism can stay in a non-equilibrium state. Usually, left to its own devices, a system comes into equilibrium with its surroundings. An unstable rock, for example, will roll until it reaches a position where the forces on it are in equilibrium – equally balanced – then it stops. Similarly, a hot cup of coffee will cool, exchanging heat with its surroundings, until the temperatures balance out. But living things stay permanently out of equilibrium. Physicists suggest that this is the definitive measure of life – a thing is living if it maintains itself in a non-equilibrium state.

Opposite left: A living organism exhibits most of the seven properties.
Opposite right: A dissipative process transfers heat to its surroundings.
Above: Phages – viruses that attack bacteria.

CRITICS

"THE EXISTENCE OF LIFE MUST BE CONSIDERED AS AN ELEMENTARY FACT THAT CAN NOT BE EXPLAINED, BUT MUST BE TAKEN AS A STARTING POINT IN BIOLOGY... THE ASSERTED IMPOSSIBILITY OF A PHYSICAL OR CHEMICAL EXPLANATION OF THE FUNCTION PECULIAR TO LIFE WOULD IN THIS SENSE BE ANALOGOUS TO THE INSUFFICIENCY OF THE MECHANICAL ANALYSIS FOR THE UNDERSTANDING OF THE STABILITY OF ATOMS."

NIELS BOHR, 1933

The Danish physicist Niels Bohr, quoted above, was central to the discovery of the quantum structure of the atom. Bohr's theory required a major shift from the previous mechanical view, which treated the electrons and nuclei of atoms as if they were simply very small particles, to accepting that quantum particles were fundamentally different in behaviour from the objects we experience. We can't deduce quantum behaviour from observing the way ordinary particles act – we have to simply accept that quantum particles are different.

Similarly, Bohr suggests, life can't be explained at the level of physics or chemistry, but should simply be taken as the way living things are. Some physicists are less comfortable with this, assuming that everything, including life, is capable of reduction to basic physical causes. In a sense this is true, but like other complex systems (for example, the Earth's weather systems) the interaction of the components is sufficiently complex that the outcome is the emergence of properties that cannot be reduced in this way.

Above: Niels Bohr (1885–1962).
Opposite top: Henrietta Lacks (1920–1951).
Opposite bottom: HeLa cervical cancer cells.

WHY IT **MATTERS**

In a sense, defining life is more a philosophical (and lexicographical) venture than a scientific one. Despite the difficulties of some specific identifications, biologists are usually confident in what it is that they study, and for the vast bulk of organisms there are no issues.

However, it can be useful to consider the nature of life – for example, in the application of the physical approach to defining life from entropy and non-equilibrium systems – as it can provide useful insights into the behaviour of other systems that are non-living but share some of these non-equilibrium properties, such as the formation of sand dunes.

There are also borderline examples that make an interesting consideration, such as the existence of cell cultures which reproduce outside of a living organism. A dramatic example is the HeLa line of human cancer cells. These cells, derived from an American cancer patient called Henrietta Lacks who died in 1951, are an important resource in cancer and HIV studies. Because cancer cells lack the controls of normal body cells they can reproduce indefinitely: over 20 tons of HeLa cells have been produced. Where it is arguable that a living cell is not alive, a cell culture like this can demonstrate most of the properties of life.

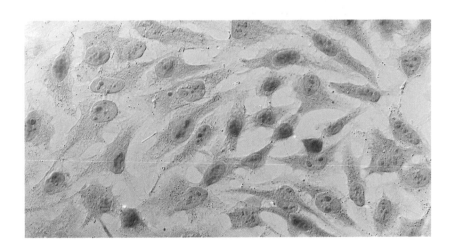

FUTURE **DEVELOPMENTS**

As there is a growing resistance to eating meat from living organisms, it has been suggested that one future approach may be to grow cells in a culture to produce meat substitutes – as with the HeLa line, it would be difficult to be entirely capable of showing whether such a cell culture was alive or not.

Perhaps the most interesting development will be in the future of artificial intelligence and artificial life. The distinction between something that simulates life (or intelligence) and is actually alive is a difficult one to make. If an artificial organism exhibits all the properties we associate with life, is it possible to definitively say that it is not alive?

We tend to make a natural assumption that such an artificial creature would not be alive, but it is arguable that this feeling is supported largely by the natural, but non-scientific, idea of a "life force". If something is artificially created, we feel it lacks the appropriate living spirit and so is not alive even if it does everything a living thing does.

Above: Meat can be grown in the lab from a cell culture.

THE **ESSENTIAL** SUMMARY

ORIGINS	KEY THEORIES AND EVIDENCE	CRITICS	WHY IT MATTERS	FUTURE DEVELOPMENTS
Historically life was associated with energy, soul or spirit. **17th century** René Descartes describes dualism with a separate "life force", mind and a body. **1701** The term "organism" is first used. **1830s** "Organism" takes on the modern usage, incorporating behaviours of a living creature. **20th century** Vitalism is confined to alternative medicine. Life is seen as an **emergent property**.	Life is usually identified by biologists by the presence of around **seven processes**. These processes include **movement, nutrition, respiration, organization, excretion, reproduction and growth**. **Not all** these processes are **strictly necessary** for life. **Viruses are a difficult case** as they lack the mechanism to reproduce, but are now more likely to be considered alive than they used to be. Physicists provide an **alternative definition of life** as a system that can reduce its entropy, maintaining itself in a non-equilibrium state.	Physicist Niels Bohr suggested that **life can't be explained** in the normal physical and chemical sense, but should be taken as the way things are, similarly to the way quantum physics has to. Others argue that life can be **reduced to atomic essentials**, but in practice understanding is limited by the **emergence of properties** from a complex system.	There is a degree to which this is just a **matter of definition**: biologists are rarely uncertain whether or not something is alive. An **understanding of non-equilibrium systems** can be useful in understanding other such systems, such as the formation of sand dunes. Borderline cases are interesting, such as the **nature of cell cultures**, which reproduce – but are they living?	There is increasing interest in **producing meat through cell cultures**. As artificial intelligence and robotics develop, the **distinction between the simulation of life and actual life** will become an important one. Doubts about an artificial creature being alive may be supported largely by the **difficulty of giving up the idea of a "life force"** behind living organisms.

PALAEONTOLOGY

THE **ESSENTIAL** IDEA

"PALAEONTOLOGY IS THE ALADDIN'S LAMP OF THE MOST DESERTED AND LIFELESS REGIONS OF THE EARTH; IT TOUCHES THE ROCKS AND THERE SPRING FORTH IN ORDERLY SUCCESSION THE MONARCHS OF THE PAST AND THE ANCIENT RIVER STREAMS AND SAVANNAHS WHEREIN THEY FLOURISHED."

ROY CHAPMAN ANDREWS, 1926

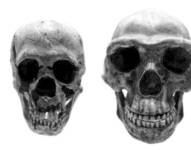

It's something of a paradox that a palaeontologist is a biologist who never has contact with living organisms. Palaeontologists are the crime scene investigators of the scientific world, examining the remains of extinct and often fossilized animals and plants to try to piece together the existence of lifeforms that are no longer present on the Earth.

The two most prominent studies of palaeontologists are the predecessors of the human species and dinosaurs. The latter gives a good example of how palaeontology has moved on in the past century. Bearing in mind the limitations of fossil remains it is remarkable how much better an idea we now have of what dinosaurs were like, from the revelation that many of these ancestors of birds had feathers (including *Jurassic Park*'s favourite, the velociraptor) to the realization that many dinosaurs were probably not cold-blooded.

Above: Ancient pre-human skulls.
Opposite top: Mary Anning (1799–1847).
Opposite bottom: Piltdown Man jawbone.

ORIGINS

Although there were some early observations of fossils – the ancient Greek philosopher Xenophanes from the fifth century BC rightly drew the inference from land-based fossil seashells that such areas were once under water, while Leonardo da Vinci illustrated a range of fossils – it wasn't until the French naturalist Georges Cuvier made a systematic comparison of fossil bones starting in the 1790s, comparing them anatomically with existing animals, that palaeontology could be considered to be a true science.

TALENTED AMATEURS

Initially, palaeontology was often the work of interested individuals outside of scientific institutions. So, for example, one of the best-known fossil collectors of the nineteenth century was Mary Anning, born in 1799, who uncovered a whole range of fossils in Dorset, England and, despite having little formal education, became a respected contributor to the field.

Originally referred to from 1776 as "fossilology", the term palaeontology was devised in 1822 by the French anatomist Henri Ducrotay de Blainville, who first tried *paléozoologie*, then broadened this to *paléosomiologie* to extend it beyond animals, but the word didn't catch on, so on his third try he ended up with *paléontologie*.

Initially, palaeontology had been an extension of comparative anatomy – looking for similarities and differences between the skeletons of animals living and long gone; with the development of evolutionary theory, it became more strongly associated with attempts to trace evolutionary trees into the past.

Despite the incorporation of evolutionary theory, palaeontology has suffered from the expectation of directed "chains of life" which run totally counter to evolutionary theory and which have inspired the concept of "missing links", resulting in some historical frauds such as the Piltdown Man, allegedly discovered in Sussex, England in 1912 as a link between apes and humans.

In practice, despite a historical tendency that continues to this day when newspapers still proclaim the discovery of new "human ancestors", the fossil record is so sketchy that it is almost impossible to put together clear trees linking humans (or many other extant organisms) to fossil remains from the distant past.

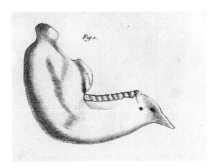
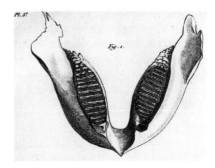

KEY **THEORIES** AND **EVIDENCE**

FOSSILS, DATING AND HUMAN ORIGIN

Central to the ability to perform palaeontology is the process of fossilization. Living things are mostly fragile, decaying over time. However, in the right circumstances, fossilization can preserve the remains of the organism indefinitely. There are a number of mechanisms, but the most important is where water that has a high mineral content seeps into remains, depositing minerals that harden to retain some of the structure of the original organism in stone.

This process is most likely to happen for sea creatures, which is why there are so many fossils found in land that was once under the ocean. For land animals the process is significantly rarer, whether we are talking about dinosaurs or early humans and the ancestor species of humans. In this regard, we are better off than the other great apes, though, as humans have often lived near the sea, while the ancestors of gorillas or chimps were more likely to have been in tropical or mountain forests, resulting in far fewer fossils.

THE TIME OF MAN

Human palaeontology gives us an estimate that our species, Homo sapiens, has existed for around 200,000 years, with older species that may be our ancestors dating back up to around 4 million years. Although the dinosaurs tend to be the poster children of truly ancient palaeontology, they only date back around 240 million years, with other lifeforms being found in fossils and lesser remains, including trace elements that seem to have originated in single-celled living organisms, dating back over 3 billion years.

Palaeontology makes use of an inverted tree structure in an approach called cladistics, which involves developing a tree showing where different species have branched off from their common ancestors.

Above: *Ichthyosaurus* fossil remains.
Opposite top: Dinosaur tree structure.
Opposite bottom: Tree ring dating used to calibrate carbon-14.

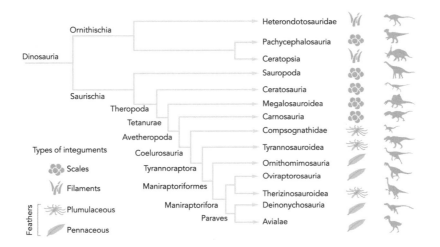

Types of integuments

- Scales
- Filaments

Feathers
- Plumulaceous
- Pennaceous

EVIDENCE

The fossil record is extremely patchy. It's not so much that there are missing links, as a missing tree, with only tiny portions of the past available to us. For example, when a pre-human fossil is found, despite the tendency of the press to label it as one of our ancestors, in practice, while we can usually place the fossil roughly in time, it isn't possible to say whether this is one of our ancestors or a species on a parallel part of the tree of life.

The ability to date fossils has benefited hugely from dating based on radioactive half-lives. As we have seen, many elements come in a number of isotopes with different numbers of neutrons in the atomic nucleus. Some of these are radioactive. When an organism is living, the radioactive atoms in it are being regularly replaced from the environment, so it has the typical levels of radioactive isotopes that were present at the time. But once the animal or plant dies, it ceases to take on new atoms, and those present gradually decay.

CARBON DATING

We can't say when any particular radioactive atom will split, giving off energy and producing a new type of atom, but what is predictable is the half-life: the time in which half of the atoms present will decay. So, for example, for an atom frequently used in dating, carbon-14, the half-life is 5,730 years. After this period has elapsed, half of the carbon-14 in a fossil remain will have decayed. After another 5,730 years half of that remaining half will have gone, and so on.

To get accurate dating this way, we need to know the proportion of carbon-14 in the atmosphere (which has varied over time). Initially this was incorrectly estimated, but it proved possible to calibrate it using the rings in very old trees (so-called dendrochronology) and it is now reasonably accurate. Carbon dating can only go so far, around 50,000 years. Other processes have to be used, for example with dinosaur skeletons, which are over 65 million years old. Often, elements such as uranium and potassium are used, where the radioactive isotopes have far longer half-lives. As dinosaurs tend not to contain uranium, usually it is the surrounding rock that is dated, rather than the dinosaurs themselves.

CRITICS

"THE PALAEONTOLOGIST, [GENETICISTS] BELIEVED, IS LIKE A MAN WHO UNDERTAKES TO STUDY THE PRINCIPLES OF THE INTERNAL COMBUSTION ENGINE BY STANDING ON A STREET CORNER AND WATCHING THE MOTOR CARS WHIZZ BY."

GEORGE GAYLORD SIMPSON, 1944

When fossils were first studied, the age of the Earth was widely accepted in the West to be around 6,000 years. This was based on biblical chronology. It was assumed that fossilized remains were of creatures which had not made it onto the ark in the great flood. However, there were some serious problems with this viewpoint. It required all the different species, extinct and still alive, to co-exist. Yet the way that fossils were found seemed to imply different timescales for different groupings of animal.

As geology developed, and with the rise of evolutionary theory, it seemed likely that life had been present on Earth for far more than 6,000 years, initially increasing the span of life to millions and then billions of years. We now believe that there has been life on Earth for as many as four billion years. This vastly expanded timescale is not accepted by relatively large numbers of people who give their interpretation of the Bible more weight than scientific evidence. They constantly search for gaps and failings in the work of palaeontology and often pick up on the big gaps in the fossil record, while ignoring the clear evidence of dating.

Above: Fossils were once thought to be remains of animals that did not survive the biblical flood.
Opposite: The mass extinction that saw most dinosaurs go extinct is thought to have been caused by an asteroid hitting Earth.

WHY IT **MATTERS**

Palaeontology is not the kind of science that has practical applications that clearly benefit everyday existence. However, it is a powerful tool for understanding the past of life on Earth, whether considering where we came from as Homo sapiens, or looking into the wider origins of existing lifeforms of all kinds.

One important outcome of the study of palaeontology is the awareness of past mass extinctions, where large swathes of the existing fauna and flora were wiped out – for example, in the mass extinction that ended the reign of the dinosaurs around 65 million years ago, thought to be linked to an asteroid impact that created devastating climate change. Some biologists believe we are in the midst of a new mass extinction at the moment as a result of human-originated changes to the environment.

"SOME BIOLOGISTS BELIEVE WE ARE IN THE MIDST OF A NEW MASS EXTINCTION AT THE MOMENT AS A RESULT OF HUMAN-ORIGINATED CHANGES TO THE ENVIRONMENT."

Palaeontology gives us a wide perspective, from the earliest known examples of life on Earth through to discoveries from relatively recent times, including remains that fit within the 200,000 years that have elapsed since our species, Homo sapiens, evolved.

As well as extending biology into the past, palaeontology gives us a better picture of the workings of evolution. For example, during the existence of human beings there have been other human species such as Homo neanderthalis and Homo floresiensis, the so-called "hobbits" of the Indonesian island of Flores, which may demonstrate an evolutionary dead end, helping to underline the lack of a single, directed chain of evolutionary development.

FUTURE **DEVELOPMENTS**

For relatively recent finds, palaeontologists are increasingly able to make use of DNA to discover relationships between species, when they diverged from each other and, for example, to discover the presence of Neanderthal genes in modern humans. However, this approach is limited, because DNA decays over time. It is unlikely ever to be possible to make use of DNA that is over 1.5 million years old – so the *Jurassic Park* scenario of recreating dinosaurs from 65-million-year-old DNA trapped in amber will never happen.

However, we are increasingly able to discover fine detail that was unavailable to earlier palaeontologists, which is why we now have, for example, some information on dinosaur pigmentation, where before colouring guesses were largely based on parallels with current reptiles. It's also the case that because fossils are so sparse, to date our picture is very sketchy, so the biggest contribution of the future will be finding more specimens – though the low probability of fossilization occurring means that our view of the past will always be limited.

Above: Some dinosaur pigmentation has now been discovered in fossil remains.

THE **ESSENTIAL** SUMMARY

ORIGINS	KEY THEORIES AND EVIDENCE	CRITICS	WHY IT MATTERS	FUTURE DEVELOPMENTS
5th century BC Ancient Greek philosopher Xenophanes inferred from sea-life fossils on land that the area was once under the sea.	Central to palaeontology is the process of **fossilization**.	When fossils were first discovered the **Earth** was widely assumed to be around **6,000 years old**.	Palaeontology does not have practical applications, but is a powerful tool for understanding the **development of life on Earth** and **where humans fit in**.	For relatively recent finds, palaeontologists can make use of **DNA to establish inter-species relationships**.
1790s French naturalist Georges Cuvier makes systematic comparison of fossil bones and modern animals.	**Sea creatures were far more likely to be fossilized** as it was more likely for mineral-rich water to seep into the remains.	The way **fossils are found in different layers** seems to imply they were not all contemporaries.	An important outcome is awareness of past **mass extinctions**, where large swathes of fauna and flora were wiped out.	DNA decays – **none is likely to exist more than 1.5 million years**, so we won't see *Jurassic Park* for real.
1820s onwards English fossil collector Mary Anning contributes many early finds.	**The fossil record** is very patchy.	We have now expanded the **timescale for life on Earth to between 3 and 4 billion years**.	Palaeontology gives us a better picture of the **working of evolution**, including the development of **other near-human species** during the time Homo sapiens has been in existence.	New fossil finds are showing more fine detail, such as some **remnants of pigmentation** from dinosaurs.
1822 The term palaeontology is coined by French anatomist Henri Ducrotay de Blainville.	**Radioactive isotope dating**, such as Carbon-14, has been widely used to discover the timescales involved.	The vastly expanded timescale is resisted by those who give their **interpretation of the Bible more weight than scientific evidence**.		The fossil **record will always be sparse**, but as we add more detail, we continue to get a **better picture of the past**.
1912 The Piltdown Man "missing link" fraud is "discovered".	**Human palaeontology** takes our species back around 200,000 years.			
	Palaeontologists use an **inverted tree structure in cladistics** to show when different species split off.			

EVOLUTION AND NATURAL SELECTION

THE **ESSENTIAL** IDEA

"EVOLUTION... IS THE MOST POWERFUL AND THE MOST
COMPREHENSIVE IDEA THAT HAS EVER ARISEN ON EARTH."

JULIAN HUXLEY, 1964

With the possible exception of climate change, there is no scientific theory that
raises as much controversy as evolution – which is odd, as it is such a common-sense
idea it is surprising it wasn't accepted much sooner. Evolution based on natural
selection takes two very simple starting points: that we pass on characteristics to
our offspring and that organisms with capabilities that enable them to survive in a
particular environment are more likely to pass on those capabilities to their offspring.

Accept these two, hardly controversial, assumptions and evolution becomes
inevitable. Because of the evolutionary process, species can adapt to changes in
their environment – and there is plenty of evidence that they do.

Those who doubt the evolutionary process often accept such adaptation on
a localized scale, but consider that it is not possible for one species to evolve
into another. However, the concept of species is a totally arbitrary label. There
is no reason why this can't happen, and a vast amount of evidence that it does.

ORIGINS

Although early ancient Greek philosophers, such as the fifth-century BC Empedocles (who devised the four-element theory), thought that it was possible for organisms to change over time, this concept was largely pushed aside by Aristotle, who set the picture of fixed species, which survived through to the Renaissance. With the 1735 *Systema Naturae* (*System of Nature*) by Swedish biologist Carl Linnaeus, more species began to be identified and structured. Linnaeus popularized the familiar binary species naming, such as Homo sapiens, but did not suggest that species could change with time.

However, by the second half of the eighteenth century, the idea that a species could change over time was in the air. For example, the English doctor Erasmus Darwin (grandfather of Charles), in a 1794 book called *Zoonomia*, suggested that all life could have originated from the same primal "filament".

THE KEY MOMENT

The key date in the history of evolution was probably 1858. After his experiences on HMS *Beagle* between 1831 and 1836, English naturalist Charles Darwin had begun the long gestation of his theory of evolution by natural selection. He seemed to have no great sense of urgency, and still hadn't published it in 1858, when he received a letter from the Welsh naturalist Alfred Russel Wallace. Wallace proposed a theory that was remarkably similar to Darwin's. Rather than race for publication, the two jointly published their ideas through the Linnean Society of London.

Darwin's highly successful book, clumsily titled *On the Origin of Species by Means of Natural Selection, or the Preservation of Favoured Races in the Struggle for Life*, arguably the best-known book in the history of science, was published on 24 November 1859 and sold out sufficiently quickly that it had to be reprinted six weeks later. Perhaps surprisingly, given later resistance to aspects of evolutionary theory, there was relatively little controversy when the book was first published.

Although there was some debate, notably the 1860 Oxford meeting, at which Samuel Wilberforce, the Bishop of Oxford, argued against the biologists Thomas Huxley and Joseph Hooker, any negative response was relatively short-lived. By the time the follow-up title *The Descent of Man, and Selection in Relation to Sex* was published in 1871, bringing humans into the evolutionary picture, the concept seemed widely accepted.

Opposite: The mudskipper is a fish that has adapted to spending time out of water.
Top: Charles Darwin (1809–1882).
Bottom: Carl Linnaeus (1707–1778).

KEY **THEORIES** AND **EVIDENCE**

THEORIES, FINCHES AND THE GENETIC SPECTRUM

Evolution is a scientific theory – but this is often interpreted as "only a theory". A scientific theory is a much stronger assertion than the way "theory" is used in general speech – the scientific equivalent is not theory, but "hypothesis". A scientific theory is an explanation backed by evidence, far more than just a hunch. In the case of evolution by natural selection, the theory is simple logic, given two assumptions. These are that parents can pass on characteristics to their offspring and that organisms better suited to their environment are more likely to breed and pass on those characteristics.

THE IMPORTANCE OF GENETICS

When evolutionary theory was first proposed, the mechanism by which the first assumption was fulfilled was not known: we are now aware that it is a matter of genetics. However, the logical implication of these assumptions, which are unquestionably supported by evidence, is that a species will change over a number of generations to reflect its environment.

Critics of evolution argue that, while "micro-evolution" involving changes within a species is possible, it doesn't explain how new species can form. However, evolution has no problem with this either, as "species" is a human, not a natural, label. A species used to be defined by the ability to reproduce with other members of its species. This definition obviously has limitations – for example, when a specific organism can't reproduce, or when a species reproduces by a single cell splitting into two new ones. Now we tend to define species genetically, which makes it clearer that the difference between two species is not an either/ or differentiator, but rather reflects a large number of small genetic differences.

Now we can understand the key paradox of evolution. Each organism is the same species as its parents – but despite this, it is possible to see how one species can eventually produce another. A useful parallel is a rainbow. Seen in detail on a computer, the colour spectrum has millions of shades. Any one of these is visually indistinguishable from the adjacent colour. But run through the entire spectrum and it shifts from red, through orange, yellow, green, blue, indigo and violet (or whatever collection of labels you would like to apply).

Similarly, the genetic variation between a parent and its offspring is tiny, making them indistinguishable as species. But over generations, when the environment makes natural selection occur, the differences can add together sufficiently to see a totally new species emerge.

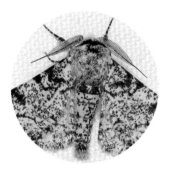

EVIDENCE

Classic examples of evolution are Darwin's finches and peppered moths. The finches have a wide variety of beak shapes and sizes. When the environment suits a big beak because, for example, weather conditions make it more likely for large, hard seeds to be available, more birds with large beaks live comfortably and produce offspring with big beaks. Similarly, when conditions tend to produce small, soft seeds, smaller beaks ensue.

The moths were originally pale with a speckled pattern that helped them blend into tree bark. With the industrial revolution, trees in some regions became much darker with soot – over a few generations, most of the moths were dark. When clean-air legislation resulted in far less pollution, after a few more generations the moths were pale again.

We have evidence both of specific new species emerging, particularly when an existing species is cut off in a different environment from its usual one, and also of the relationship between species in the amount of their DNA that is shared. It's rare that we get to see species change because of the large timescales involved. Life has existed on Earth for around 4 billion years – a lot of time for variation to occur.

Opposite: Darwin's diagram of finch beaks, evolved to suit the environment.
Top: A pale peppered moth.
Bottom: One of the range of Darwin's finch species.

CRITICS

"EVOLUTION EVER CLIMBING AFTER SOME IDEAL GOOD, AND REVERSION EVER DRAGGING EVOLUTION IN THE MUD."

ALFRED, LORD TENNYSON, 1886

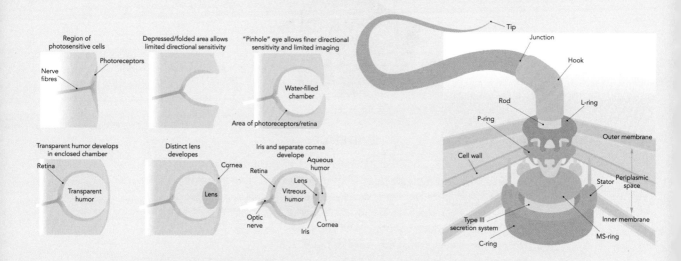

After its initial surprisingly easy start, evolution has faced significant opposition, often from religious groups who believe that it runs counter to, for example, scriptural assertions that humans and animals were created at the beginning of the planet in the forms they have today. Although many with a religious belief are happy to accept evolution as the mechanism used by their deity, others feel it runs counter to their beliefs.

CREATIONISM

Although creationists simply refuse to accept evolution as they believe their religious texts override any scientific observations, others – notably supporters of the concept known as intelligent design – suggest that evolution can be countered by establishing examples, such as the eye, or the motor-like mechanism of the bacterial flagellum, which they argue could not have evolved because intermediate stages on the way to their development had no evolutionary value.

This argument suggests that small-scale evolution is possible, but not major changes or developments. However, the argument fails, sometimes because there are clear benefits from intermediate stages – there are types of eyes, for example, in a whole range of increasingly sophisticated mechanisms – or because potential intermediate stages can have other benefits, or simply be tolerated because another aspect of the organism is very successful.

WHY IT **MATTERS**

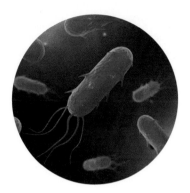

Evolutionary theory is the central tenet of biology. Before the theory was developed it was arguable that biology was primarily a matter of observation and cataloguing with very little science involved – hence physicist Ernest Rutherford's infamous jibe, "All science is either physics or stamp collecting." But evolution put biology on a structured scientific basis: like all good scientific theories, it was adding explanation to information.

"EVOLUTION EXPLAINS HOW WE, AND ALL THE EXISTING ORGANISMS, CAME TO HAVE THE DIVERSITY WE EXPERIENCE TODAY."

Evolution explains how we, and all the existing organisms, came to have the diversity we experience today. Although genetics would be required to provide an appropriate mechanism for this to occur (see the next section), with evolutionary theory in place it was finally possible to make sense of the whole panoply of life, from single-celled organisms such as bacteria all the way through to complex animal and plant lifeforms.

Evolution also matters hugely as a triumph of logic over received wisdom. It reflects the motto of Britain's Royal Society, one of the world's oldest scientific societies founded in 1660. The society's motto is the Latin *Nullius in verba*, roughly translated as "Take no one's word for it." As this suggests, our scientific theories should be based on evidence, rather than what we are told to think. Evolution is entirely based on evidence, rather than received wisdom.

Opposite left: Different stages of evolution of an eye, all functional.
Opposite right: The remarkable molecular machinery of the bacterial flagellum.
Above: Flagellate bacteria.

FUTURE **DEVELOPMENTS**

The basic elements of evolution by natural selection can never be countered because they are purely logical. However, the second motto of scientists after *Nullius in verba* should be: "It's more complicated than we thought." Although the process of natural selection is always at work, there are other elements that had to be added in, because the realities of nature are more complicated.

Darwin himself would add in the concept of sexual selection, where a development may make a species less fit to survive in terms of avoiding predation, for example, but better able to breed due to making the individual more attractive to the opposite sex. A clear example is the peacock's tail, which surely makes it less likely to survive both in its visibility and clumsiness, but is outweighed by its attraction for peahens.

New factors continue to be added to the evolutionary picture, such as the role played by epigenetics – the impact of mechanisms that control the operation of DNA and that allow, for example, for some environmental factors that had an effect on the parents influencing the development of the offspring.

Above: The peacock's tail counters reduced survival with increased attraction.

THE **ESSENTIAL** SUMMARY

ORIGINS	KEY THEORIES AND EVIDENCE	CRITICS	WHY IT MATTERS	FUTURE DEVELOPMENTS
5th century BC Ancient Greek philosopher Empedocles states that organisms can change over time.	Evolution by natural selection is a **logical, evidence-based theory**.	Critics of evolution often do so as it **runs counter to ancient religious texts**. This is not a logical argument and can't be handled scientifically.	Evolutionary theory is a **central tenet of biology**, adding explanation to information.	The basics of evolution can't be countered, but **it is more complicated** than first thought.
4th century BC Ancient Greek philosopher Aristotle suppresses the theory of species change.	Evolution is based on two assumptions: **parents can pass on characteristics to offspring**, and **organisms with characteristics enabling them better to survive are more likely to reproduce**.	Some critics of evolution use the concept of **intelligent design**, suggesting that there are **some features of living organisms that could not have evolved** because intermediate stages had no value.	Evolution explains the **origin and diversity** of species.	Darwin added the concept of **sexual selection**, which can result in evolutionary changes that make life more risky.
1794 English naturalist Erasmus Darwin suggests all life could have originated from the same primal "filament".	Classic examples of small-scale evolution are **Darwin's finches** and the **peppered moth**.	This argument fails in some cases because **there are clear examples of intermediate stages with value** (e.g. the eye).	Evolution is a triumph of **logic based on evidence** over received wisdom.	**New factors** continue to be added, such as **epigenetics, the mechanism by which DNA is controlled**, which allows some environmental influences on parents to be passed on to offspring.
1858 Charles Darwin and Wallace publish the theory of evolution by natural selection.	Each **organism is the same species as its parent**.	In other cases, the intelligent design argument fails because **intermediate stages have alternative benefits or are tolerated because of other benefits** (e.g. bacterial flagellum).		
1859 Charles Darwin's *On the Origin of Species* is published.	Because **small changes accumulate from generation to generation**, eventually new species emerge.			
1871 Charles Darwin's *The Descent of Man* is published.	New **species have been observed**, and **all known species are genetically linked**.			

GENETICS

THE **ESSENTIAL** IDEA

"GENETICS WAS THE FIRST BIOLOGICAL SCIENCE WHICH GOT IN THE POSITION IN WHICH PHYSICS HAD BEEN IN FOR MANY YEARS."

THEODOSIUS DOBZHANSKY, 1962

When Darwin came up with his theory of evolution, he was missing an essential piece of the jigsaw in understanding how traits were passed on from one generation to another. The first steps in genetics were already being taken, but would not be widely known until later.

Genetics is based on the functions of DNA, the extremely long biological molecules used to store the information that is passed from one generation of an organism to another, carrying with them instructions for creating essential biological molecules, notably proteins.

Despite the way they are sometimes portrayed, the genes in DNA are rarely individually responsible for a particular aspect of an organism. It's more common for characteristics to be based on a combination of a range of different genes.

There is also a large input to the development of an individual from the environment. This is both true at the simplistic level of, for example, having sufficient nutrition, to the more sophisticated impact of epigenetics, where environmental factors can result in different genes being switched on and off.

Above: DNA stores information passed on from one generation to the next.
Opposite left: Mendel experimented with pea plants, comparing inherited characteristics.
Opposite middle: Gregor Mendel (1822–1884)
Opposite right: Part of Crick and Watson's original DNA model.

ORIGINS

  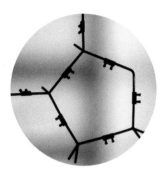

As long as humans have wondered why offspring resemble their parents there has been an interest in "heritance" – but genetics supplied the mechanism that supported that process.

Over ten years from the mid-1850s, Austrian friar Gregor Mendel (born in what is now the Czech Republic) experimented with pea plants, looking at characteristics such as height, colour, seed shape and so forth. He inferred from his results that there were specific factors, coming from the male and female organisms in one generation, which influenced the outcome, depending on whether a trait had to come from one or both parents to be passed on.

This established the process, but not the mechanism. The biological molecules that would become known as DNA were identified in 1878 by Swiss physician Friedrich Miescher, but DNA's function was not known. In 1902, American physician Walter Sutton and German biologist Theodor Boveri both identified large molecular strands in cells known as chromosomes as the likely vehicle for transmitting heritance. By 1922, American zoologist Hermann Muller was able to publish a paper entitled "Variation due to the Change in the Individual Gene". While having no idea of what genes *were*, he was able to describe them as "ultramicroscopic particles" which played a fundamental role in determining cell substances, structures and activities.

THE GENETIC REVOLUTION

Genes were identified in fruit flies as being parts of chromosomes, though Muller observed: "the chemical composition of the genes… remain as yet quite unknown." By 1943, American medical researcher Oswald Avery suggested from experiments on bacteria that DNA was involved in carrying information between organisms – chromosomes would prove to be single DNA molecules.

Also in 1943, Austrian quantum physicist Erwin Schrödinger gave a series of lectures in Dublin, published in the book *What is Life?* in 1944. Amongst other things, Schrödinger suggested that the hereditary mechanism would be based on what he called an "aperiodic crystal". This is a substance that doesn't have the regular structure of a typical crystal, but rather can store information along its length: this perfectly describes DNA.

The final part of laying the foundations of genetics came in the discovery of the structure of DNA in 1953 in Cambridge by James Watson, Frances Crick, Rosalind Franklin and Maurice Wilkins, while the discovery of the mechanism used to code for the different amino acids used to build proteins would follow in the early 1960s.

KEY **THEORIES** AND **EVIDENCE**

DNA, GENES AND CHARACTERISTICS

The cells of practically all organisms contain chemical molecules known as DNA – deoxyribonucleic acid. This is not a single substance, but a family of near-infinite variations. DNA consists of two helical strands of linked sugars which are joined together at regular intervals by groups of atoms known as base pairs. The whole looks something like a spiral staircase with the base pairs forming the treads.

Living cells typically contain a number of strands of DNA called chromosomes, each a DNA molecule. The base pairs in the strand are in effect an information store: each pair can be based on four different compounds: cytosine, guanine, adenine and thymine. In effect, these four are the "letters" making up the genetic code. What is clever about the DNA structure is that it has a built-in mechanism to duplicate that information.

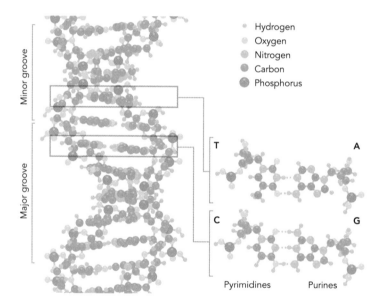

Each base pair consists of two compounds, which always pair the same way: cytosine with guanine and adenine with thymine. So, it is possible to unzip a strand of DNA, splitting it in two down the middle of each base pair. Knowing which base is present on one strand enables a new strand to be built, because of this standard pairing approach.

The primary genetic information stored in DNA is in the form of genes – collections of "codons", which are triplets of base pairs that carry the code to build one of twenty amino acids, the essential building blocks of life. However, it's not possible to look at a stretch of DNA and see the genes. The codons that make up a gene are interspersed with other information, so the cellular mechanism that reads DNA has to first cut out various parts of the information and splice the relevant parts back together before taking action.

In the case of human DNA, only around 2 per cent forms the genes. Much of the rest was once labelled as junk, but much of this is now known to be responsible for controlling the way that the genes are switched on and off, so forms part of the larger "program" for life.

Some forms of life reproduce by splitting cells from a single parent, producing offspring that are clones of the original with near-identical genes. But most organisms make use of sexual reproduction, where the genes of the offspring are made up of a mix of components from both of the parents. This is a valuable process because it provides the differences between individuals necessary for evolution through natural selection.

GENE TRANSFER

Where an organism does reproduce by splitting, it often makes use of an alternative approach to ensure that there is still variation. So, for example, bacteria employ a process called horizontal gene transfer, where genes can be swapped between individuals. This is why bacteria are able, for example, to pass on antibiotic resistance, which will have developed as a mutation in one organism, but then gets passed on to others.

Mutation is an important factor in genetics. This is when one or more of an offspring's genes are different from those that were passed on from the parent. The most common ways mutations happen are when the mechanism reproducing the genes makes a copying error, or when radiation causes a change in a DNA molecule. Each human, for example, will typically have a few million tiny mutations, most of which will have no impact, but a few of which may prove beneficial or problematic.

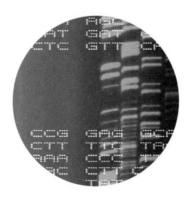

EVIDENCE

The evidence for genetic theory was initially inference from the way different characteristics were passed on, but over time we have developed better mechanisms for examining and comparing DNA and for observing the ways that the molecular machinery of cells operate.

Opposite: The dual helix structure of DNA.
Above: Codons, each represented by three letters, carry the genetic code.

CRITICS

"IN A SENSE, GENETICS GREW UP AS AN ORPHAN. IN THE BEGINNING BOTANISTS AND ZOOLOGISTS WERE OFTEN INDIFFERENT AND SOMETIMES HOSTILE TOWARDS IT."

GEORGE WELLS BEADLE, 1958

Historian of science Thomas Kuhn described the concept of a paradigm shift, where the view of a particular area of science undergoes a revolutionary change, rather than the usual small incremental changes. Genetics was such a shift for biology, turning the whole thing on its head and putting the molecule DNA and its structure in charge, a viewpoint typified by English zoologist Richard Dawkins's famous book *The Selfish Gene*.

As is always the case in a scientific revolution, some were left behind and others opposed the new, often while not understanding it. Apart from criticism of the mechanism itself, much opposition arose from the misuse of genetics, both as theory and practice.

EUGENICS

The theoretical problem arose from the work of Darwin's cousin, English polymath Francis Galton. It was Galton who in the 1880s came up with the term eugenics to describe the potential of using genetic information to breed a "better" human – or prevent breeding of "lesser" humans. Eugenics has been used to justify appalling acts, such as sterilizing those considered inferior to prevent reproduction. Eugenics has no scientific basis.

The practical issue comes from genetic modification. We now have mechanisms, for example with the gene-editing tool CRISPR, to modify DNA. While this is rightly treated with considerable caution, especially when used on animals, it can make significant improvements in crops and has the potential to be extremely beneficial if correctly controlled, but has produced knee-jerk criticism, such as the European Union's ban of genetically modified organisms.

Above: Francis Galton (1822–1911).
Opposite: Part of the human genome map.

WHY IT **MATTERS**

Genetics explains how the biochemistry of living things shapes their nature – if evolution was the first truly scientific aspect of biology, genetics provides the mechanism for that science.

The importance of genetics is hard to overstate, though it can lead to misunderstandings. When the human genome – a near-complete list of the genes in a human (actually a mix from a number of people) – was first published there was much excitement about the ability, for example, to develop targeted medication that would be tailored to an individual's genes, or to provide gene therapy to deal with a medical condition, but it has proved more complicated than first thought.

> "A MAJOR ISSUE IS THE COMPLEXITY OF THE RELATIONSHIP
> BETWEEN GENES AND SPECIFIC OUTCOMES."

A major issue is the complexity of the relationship between genes and specific outcomes. Although there are a few traits (for example, red hair) and medical conditions (such as cystic fibrosis) tied to a single gene, more often than not it is the interplay between a number of genes that results in a particular outcome.

This doesn't in any way take away from the benefits of genetics – it is still the fundamental information mechanism for transfer of inherited characteristics between generations, and is responsible for everything from the structures of all living organisms to at least half of human behaviour. Just as computers make use of a binary code of zeroes and ones, DNA means that living things share information in a quaternary code of base pairs.

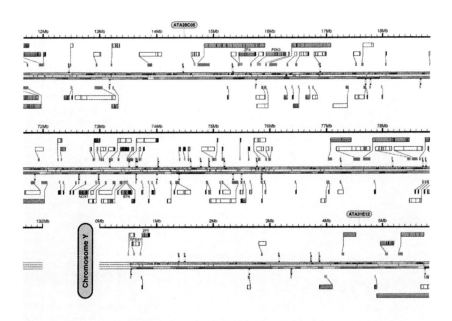

FUTURE **DEVELOPMENTS**

The basics of genetics are well understood, but the wider science of epigenetics is still relatively new and continues to be developed.

Although there is a clear link of genes to the production of amino acids, which are used to build proteins – and the functioning of these genes contributes to many of the characteristics that define the nature of a particular organism – epigenetics refers to mechanisms that occur outside of the genes. The control mechanism for genes is in the DNA that doesn't code for genes – in the case of humans, 98 per cent of DNA is non-coding. This can include, for example, mechanisms for methylation, where small atomic groups are added to the DNA, resulting in specific genes being turned on and off at various times.

Because epigenetics is not "locked in" to the genes, but can vary during an organism's lifetime, the environment can have a direct impact on the epigenetic content, which can even be passed on to offspring.

POLYGENIC SCORES

The latest frontier is polygenes or multiple-gene inheritance. Behavioural geneticists are rapidly coming to understand that human differences are very rarely the result of single genes. They are usually caused by multiple genes, each one of which has a tiny influence, but which can collectively cause statistically significant variations when viewed across populations and, increasingly, even at the level of individuals. The discipline is in its infancy, but "polygenic scores" can already predict a remarkably wide variety of traits, from body weight to academic performance and even propensity for watching television.

Although the initial promise of gene-based tailored medication was overplayed, as the relationship between genes and health is complicated, nonetheless we can expect to see major developments in molecular biomedical science in the future.

Above: Methylation sites on DNA where groups can be added.

THE **ESSENTIAL** SUMMARY

ORIGINS	KEY THEORIES AND EVIDENCE	CRITICS	WHY IT MATTERS	FUTURE DEVELOPMENTS
1850s Gregor Mendel experiments with breeding pea plants and infers the existence of factors passed from generation to generation. **1878** DNA discovered by Friedrich Miescher. **1902** Walter Sutton and Theodor Boveri separately identify chromosomes as likely vehicles for transmitting inheritance. **1922** Hermann Muller describes "variation due to the change in the individual gene". **1943** Oswald Avery suggests DNA is the molecule carrying information and Erwin Schrödinger predicts an aperiodic crystal will fulfil the role. **1953** James Watson, Frances Crick, Rosalind Franklin and Maurice Wilkins identify the structure of DNA.	**Almost all biological cells contain DNA**, a complex molecule which contains information in its structure. Information in DNA is **stored in base pairs** of the compounds cytosine, guanine, adenine and thymine. DNA is structured so it can "unzip" down the middle and **duplicate its information** when cells split. The primary **information in DNA comes in the form of genes**: collections of base pairs which specify the structure of protein molecules. In sexual reproduction, **genes of an offspring are made up of a mix of components** from the parents, providing variation. **In a mutation, an offspring's gene is changed from the inherited form**, typically by a copying error or radiation.	Genetics was a **revolutionary change** to the approach to biology, resulting in some natural resistance. There has been criticism of the theoretical **misuse of genetics in eugenics**. This attempt to breed "better" humans **relies on a non-scientific assumption** that it is possible to identify which beneficial characteristics an offspring will have from its parents. By contrast, genetic modification is good science, and if carefully controlled can make significant advances in crops etc., but is often resisted without a scientific basis.	Genetics explains how the **biochemistry of living things shapes their nature**. The publication of the **human genome** produced considerable hype about genetic tailored **medication and gene therapy**. This is not without value, but is much harder to make use of than first thought. Genetics remains the fundamental **information mechanism for the transfer of inherited characteristics**. Just as computers make use of a binary code of zeroes and ones, DNA means that **living things share information in a quaternary code** of base pairs.	The basics of genetics are well understood, but there is still **much to learn about epigenetics**. Although the initial promise of gene-based tailored medication was overplayed, we can **expect to see major developments in molecular biomedical science** in the future. **Polygenic scores** are the latest frontier in behavioural genetics, allowing the discipline to move beyond simplistic notions of single-gene causality.

SIMPLE LIFE FORMS

THE **ESSENTIAL** IDEA

"BUT HOWEVER SECURE AND WELL-REGULATED CIVILIZED LIFE MAY BECOME, BACTERIA, PROTOZOA, VIRUSES, INFECTED FLEAS, LICE, TICKS, MOSQUITOES AND BEDBUGS WILL ALWAYS LURK IN THE SHADOWS..."

HANS ZINSSER, 1934

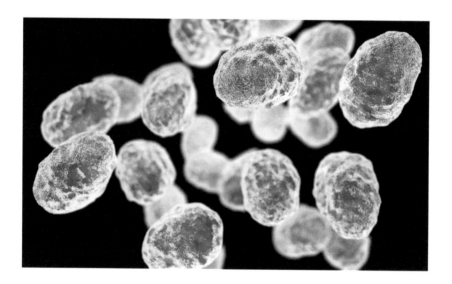

Life appears to have first arisen on Earth around 4 billion years ago, and for the first half of that time, the only organisms were the simple single-celled lifeforms known as prokaryotes: bacteria and archaea.

Although they were later joined by eukaryotes, organisms that could have one or more cells with significantly more complex structures (including, eventually, humans), this doesn't mean that the prokaryotes lost their grip. They still numerically dominate life on Earth. As a simple example, humans on average are thought to have as many prokaryotes in their bodies as they have their own cells – perhaps 30 trillion. We are seriously outnumbered.

Although there is still dispute over whether or not viruses are alive, they can't be ignored and fit better here than anywhere else. The exact origins of viruses are not clear, but there is molecular evidence that viruses predate the last common ancestor of all living things, so certainly date back over 2 billion years and are likely to have come into existence with the earliest lifeforms. We tend to think of viruses as an affliction of animals, but the majority attack bacteria and archaea, making prey available from the very beginning of life.

ORIGINS

The existence of tiny, unseen lifeforms was speculated on since ancient times, but arguably there was no scientific evidence until the mid-seventeenth century. German polymath Athanasius Kircher observed what he described as "animalcules" in blood from victims of the plague in 1658, though there is some suspicion that his simple microscope did not have the magnification to see the relatively small Bacillus pestis responsible for the plague, and it is likely he saw blood cells.

What is more certain is that Dutch draper Antonie van Leeuwenhoek discovered more sizeable Selenomonas bacteria in 1683, several years after spotting other microscopic organisms. Although compound microscopes had been available for several decades, his discovery was the result of using an extremely simple single-lensed microscope.

Bacteria would be given a classification of their own by German biologist Ferdinand Cohn in the 1870s. Another group of prokaryotes, archaea, were first thought to be a type of bacteria, but when it became possible to study them at the genetic level, they proved to be more similar to a eukaryotic cell (though without a nucleus) than a bacterium and were split off as a totally separate domain in 1977.

VIRUSES

Viruses were not discovered until significantly later than bacteria as they are mostly very small (though we now know that there are a few unusually large viruses). A virus is distinct from all other forms of life because it does not have the means to reproduce – instead it hijacks the cell-reproduction mechanism of its host, destroying host cells in the process. The existence of viruses was speculated on before they were discovered. It is difficult to pin down a first discoverer of them. Perhaps the most likely was Dutch chemist Martinus Beijerinck in 1898, building on the work of Adolf Mayer and Dmitri Ivanowsky, who first clearly showed that viruses existed, while bacteriophages (viruses that attack bacteria) were first found by English bacteriologist Frederick Twort in 1915.

Opposite: Tularaemia bacteria.
Top: Martinus Beijerinck (1851–1931).
Bottom: Replica of a van Leeuwenhoek microscope.

KEY **THEORIES** AND **EVIDENCE**

MICROBES, CELLS AND (NO) NUCLEI

"[VIRUSES] CAN BE ANALYZED AND PARTLY UNDERSTOOD ACCORDING TO THE RULES OF A SIMPLER DISCIPLINE, BUT THEY ALSO PRESENT ANOTHER LEVEL OF COMPLEXITY: VIRUSES ARE LIVING CHEMICALS…"

ALISON JOLLY, 1985

Bacteria and archaea are single-celled organisms known as prokaryotes. They are in effect a bag formed by the cell wall containing chemicals, DNA and molecular machinery. Unlike the other main type of organism, eukaryotes, prokaryotes have no nucleus, which is an inner section of the cell separated off by a membrane containing the vast majority of a eukaryote's DNA. Lacking a nucleus, prokaryote DNA is found within the main cell membrane. Rather than being stored coiled up around a structure called a histone like eukaryote chromosomes, the single prokaryotic chromosome has a more tangled appearance and is usually circular.

Prokaryotes reproduce asexually through binary fission: a single cell divides into two identical cells. This process begins by splitting the DNA and constructing two complete copies; the two chromosomes are moved to opposite ends of the cell and then the cell divides down the middle. As mentioned above, although the basic means of reproduction does not offer the opportunity to mix genetic material from sexual reproduction, genes are passed from organism to organism through the lateral transfer mechanism.

MEANS OF PROPULSION

A number of bacteria and archaea are equipped with an external structure called a flagellum. This is primarily a means of propulsion. The flagellum (named from the Latin for "whip") spins around, operating like a tiny propeller. Remarkably, to power the flagellum, the prokaryotes have a rotary molecular motor. This is electrochemically powered by the movement of positively charged proton particles and typically rotates at up to a thousand revolutions per minute. Most prokaryotes are between 500 and 5,000 nanometres across (a nanometre is a billionth of a metre – about 40 billionths of an inch). By no means all are pathogenic. Those that are can cause damage by directly attacking host cells, by producing toxins or by stimulating an excessive immune response. Friendly bacteria exist in a wide range of locations in the body, but are particularly concentrated in the gut, where they help break down difficult-to-digest materials, and make use of the appendix as a safe location to breed, away from the stressful acidic environment of the main digestive system.

Although prokaryotes come in a range of shapes and sizes, they are less varied than viruses, which can have bizarre

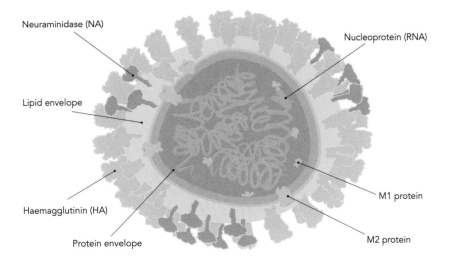

Neuraminidase (NA)

Nucleoprotein (RNA)

Lipid envelope

Haemagglutinin (HA)

Protein envelope

M1 protein

M2 protein

shapes, from phages that resemble moon landers to rotaviruses looking like a traditional spiked sea mine. Despite some still considering viruses not to be living, they have many of the characteristics of a living organism, such as genes and the ability to evolve through natural selection. However, they lack a cell wall, instead having a structure based on a shell of protein called a capsid, and they cannot reproduce without making use of the mechanism of a prokaryotic or eukaryotic cell.

GOING VIRAL

Viruses reproduce by assembling copies of themselves. They tend to be significantly smaller than prokaryotes – typically 20 to 300 nanometres in size – though one type of virus, the mimivirus, which attacks amoeba, is close to bacterial size at around 400 nanometres across, and was mistaken for a bacterium when first discovered.

Although some viruses, notably phages, have their genes in the form of DNA, many instead use a related structure known as RNA (ribonucleic acid). RNA is usually single-stranded, unlike DNA's double helix, and uses one different base (uracil rather than thymine). RNA has an important function in all living things, as it is used in the process that reads information off DNA to specify amino acids to build proteins.

EVIDENCE

Our understanding of the structure of prokaryotic cells and viruses has grown with inventions for investigating tiny structures, from microscopes to electron microscopes, X-ray diffraction and sophisticated scanners. The relationships between species is, to some extent, derived from tracing parallel segments of DNA or RNA that can be found in different organisms.

Opposite top: Typical prokaryote structure.
Opposite bottom: Mats formed by colonies of archaea.
Above: Influenza virus structure.

CRITICS

The widest criticism of the science of bacteria and viruses has tended to come from the "anti-vaccination" movement or "anti-vaxxers", who believe that vaccines (which prompt the body's immune system to defend against attack) are unsafe. They claim that vaccines cause a range of problems for individuals, from inducing autism to weakening the immune system. In some countries, conspiracy theories have linked vaccination to attempts by foreign powers to render particular ethnic or religious groups infertile.

THE MMR SCANDAL

The best example of the malign influence of the movement is the success of the now-struck-off medic Andrew Wakefield in persuading large numbers of parents that the MMR vaccine (combined measles, mumps and rubella) caused autism in children – a claim that lacked evidence and has been disproved using large-scale trials. As a result of the ensuing media scare, and support for the anti-vaccine movement from a number of celebrities and social media influencers, sufficient numbers of parents in Europe and the US prevented their children from having the vaccination for numbers to drop below the "herd immunity" level, where a sufficient percentage of the population are protected to prevent the disease spreading. The result has been repeated outbreaks of the dangerous childhood disease measles, leading to serious illness and deaths.

WHY IT **MATTERS**

There are two reasons that prokaryotes and viruses are hugely important. These simple organisms were the only forms of life on Earth for around 2 billion years, and all life – including eventually humans – has evolved from early forms of bacteria, archaea or a combination of the two. These organisms represent the beginnings of life on Earth – a hugely important scientific development.

The other, more selfish significance is their impact on human life. Beneficial bacteria are vital to the effective functioning of our digestive systems, while pathogenic bacteria are responsible for a wide range of diseases, including tuberculosis, pneumonia, tetanus, typhoid, diphtheria and a range of bacterial food poisoning.

Viruses are also responsible for many afflictions, from the relatively mild, such as the common cold, through to life-threatening diseases such as influenza, hepatitis, ebola, AIDS, rabies, SARS and COVID-19.

Like bacteria, viruses also have their positive side from the human viewpoint. The vast majority of viruses attack bacteria. Such bacteriophages (or phages) were seen early on as a potential treatment for bacterial infection, but the approach was largely dropped with the discovery of antibiotics. However, in the twenty-first century, as some dangerous bacteria develop resistance to antibiotics, meaning that they become near-impossible to wipe out this way, there is a renewed interest in the possibility of using phages for their medical benefits.

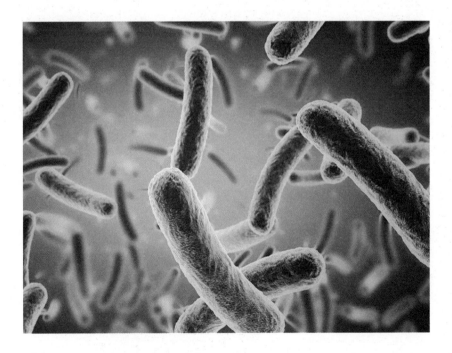

Opposite top: Measles rash.
Opposite bottom: Insufficient vaccination leads to loss of herd immunity.
Above: Bacteria, both friendly and pathogenic, exist in vast numbers.

FUTURE **DEVELOPMENTS**

Much of the work on bacteria and viruses is inevitably medical, looking for ways to reduce the impact of pathogenic microbes. When attempting to control bacteria, as well as the increased focus on phages, there has been some effort on developing new antibiotics. For decades this research had stalled, as there were limited financial rewards for pharmaceutical companies in the huge investment required in investigating new drugs.

However, in 2020, a new approach, using artificial intelligence, was developed to take on the job of searching through millions of chemical structures to find ones that were likely to be able to attack the structures of bacteria in new ways. This has already resulted in one powerful new antibiotic being discovered and many more could be found this way.

Viruses are not susceptible to antibiotics, so the dual line of attack for the future is the development of new vaccines, which can stimulate the body's defences into taking on a particular virus, and the production of new antiviral drugs, which act to prevent viruses from developing within the host.

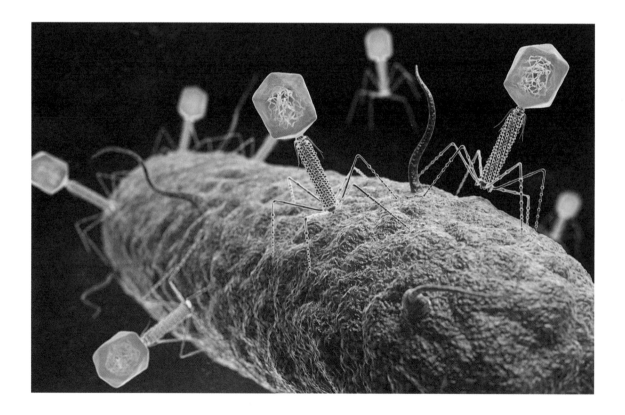

Above: Viruses that attack bacteria known as phages could help replace some failing antibiotics.

THE **ESSENTIAL** SUMMARY

ORIGINS	KEY THEORIES AND EVIDENCE	CRITICS	WHY IT MATTERS	FUTURE DEVELOPMENTS
1658 Athanasius Kircher observed "animalcules" in plague victims' blood, but these were probably cells. **1683** Antonie van Leeuwenhoek discovers Selenomonas bacteria. **1870s** Bacteria given classification by Ferdinand Cohn. **1898** Martinus Beijerinck clearly shows viruses exist. **1915** Frederick Twort identifies bacteriophages. **1977** Archaea are split off from bacteria as a separate domain of life.	Bacteria and archaea are **prokaryotes**, single-celled organisms without a central nucleus. The DNA of prokaryotes is usually in a **single, tangled-appearing, circular chromosome**. Prokaryotes **reproduce asexually** by cell division. A number of prokaryotes have a **propeller-like flagellum**, powered by a molecular motor. Prokaryotes are typically **500 to 5,000 nanometres across**, where viruses are typically **20 to 300 nanometres**. Viruses **reproduce using the mechanisms of their host cells**, assembling copies rather than dividing. Although some viruses use DNA, **most have their genes in the form of RNA**.	The widest criticism of the science comes from the **anti-vaccination movement**. Anti-vaxxers believe that **vaccines are unsafe**, for example causing autism in children. Former doctor Andrew Wakefield **persuaded many parents that the MMR vaccine caused autism**. As a result, **a number of countries no longer have "herd immunity"** for measles, leading to serious illness and deaths.	These simple organisms were the **only forms of life on Earth for 2 billion years** and are our distant ancestors. Understanding both the **potential benefits and diseases** these organisms cause is extremely important for human health. **Some bacteria are beneficial**, helping our digestive system. The **majority of viruses (bacteriophages) attack bacteria**, and could be helpful in the face of bacterial antibiotic resistance.	After years of stagnation, **artificial intelligence is starting to be used to help develop new antibiotics**. **Phages may be of increasing use** in medical applications. **New vaccines and antiviral drugs** will continue to be needed.

COMPLEX LIFE FORMS

THE **ESSENTIAL** IDEA

"THUS A EUKARYOTIC CELL MAY BE THOUGHT OF AS AN EMPIRE, DIRECTED BY A REPUBLIC OF SOVEREIGN CHROMOSOMES IN THE NUCLEUS."

GUNTHER STENT, 1971

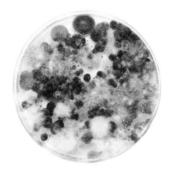

As we have seen, organisms are primarily divided into those based on prokaryotic and eukaryotic cells. Eukaryotic organisms have the more complex type of cell with a nucleus surrounded by a membrane which contains the chromosomes and a range of other structures outside the nucleus, including the "power source" units called mitochondria.

The eukaryotes are far more variable in form, from single-celled organisms such as amoeba and paramecia through to human beings. Eukaryotes are broadly grouped into four or five "kingdoms". The easily discerned three are animals, plants and fungi. The remainder are either lumped together as protists, or divided into chromista (which have structures called plastids using chlorophyll to generate energy from light, and/or specific types of the hair-like structures for movement called cilia) and protozoa, which include amoebas and plasmodia, including the malaria parasite.

The term "eukaryote" comes from the Greek for "true kernel" – a reference to its nucleus (prokaryote derives from "before kernel").

ORIGINS

Where it only became possible to know of the existence of prokaryotes once microscopes were available, we have always been aware of eukaryotes, making up as they do all the familiar animals, plants and fungi that surround us on the planet (and humans ourselves). As eukaryotic cells are significantly bigger than prokaryotes – typically around twenty times larger – their discovery took less time, though even these cells mostly need microscopes to be seen in any detail. English scientist Robert Hooke came up with the term "cell" when studying layers of cork under a microscope for his 1666 book *Micrographia*, naming them after monastic cells.

"ROBERT HOOKE CAME UP WITH THE TERM 'CELL' WHEN STUDYING LAYERS OF CORK UNDER A MICROSCOPE FOR HIS 1666 BOOK *MICROGRAPHIA*, NAMING THEM AFTER MONASTIC CELLS."

Over time, different components of the eukaryotic cell were identified. The nucleus may have been spotted by Van Leeuwenhoek in the early eighteenth century, and was more clearly described by Scottish botanist Robert Brown in 1831. By the 1860s, the existence of chromosomes had been suggested; mitochondria were identified by German pathologist Richard Altmann in 1890, while the gradual understanding of their function was developed between 1925 and 1948.

The specific distinction between eukaryotes and prokaryotes was made by French biologist Édouard Chatton in 1925, and developed in 1962 by Canadian biologist Roger Stanier and Dutch-American biologist Cornelius van Niel.

The idea that mitochondria were once independent bacteria was proposed by American biologist Lynn Margulis in 1966, though not widely accepted until the 1980s. The final important piece of the eukaryotic jigsaw was added in 1977, when "archaebacteria", later renamed archaea, were added as a distinct domain of life from bacteria. It is now thought that early eukaryotes were formed when archaea absorbed bacteria to form a compound organism.

Opposite: Eukaryotes include moulds, part of the fungus kingdom.
Top: Lynn Margulis (1938–2011)
Bottom: Hooke's drawing of cork cells.

KEY **THEORIES** AND **EVIDENCE**

CELLS, NUCLEI AND MORE

"THE NUCLEUS HAS TO TAKE CARE OF THE INHERITANCE OF THE HERITABLE CHARACTERS, WHILE THE SURROUNDING CYTOPLASM IS CONCERNED WITH THE ACCOMMODATION OR ADAPTATION TO THE ENVIRONMENT."

ERNST HAECKEL, 1866

Eukaryotic cells are literally defined as having a nucleus. Note that this is a *typical* cell in a eukaryotic organism – some cells, for example red blood cells, may not contain nuclei.

As we have seen, eukaryotic cells often also contain other smaller structures known as organelles. The best known of these, mitochondria, are responsible for reacting fuel molecules, derived from food, with oxygen, transported by red blood cells. The mitochondria produce a molecule called adenosine triphosphate (ATP), which is used as an energy store, transported around the body to locations where it is required. This is such an intensive system that, for example, humans produce about their own bodyweight in ATP each day.

PHOTOSYNTHESIS

In plants, algae and chromista, another common type of organelle is the chloroplast, where photosynthesis takes place. These tiny structures (there can be as many as 100 of them in a single plant cell) contain extremely complex molecular structures to carry out photosynthesis, the power source of arguably the majority of life on Earth. Most organisms take their energy either directly from the Sun or through eating organisms that have absorbed energy this way.

In the photosynthesis process, a photon of light provides energy to boost the energy of an electron in a molecule of chlorophyll. In a quantum process, this energy is passed from molecule to molecule until it reaches a part of the chloroplast called the photosynthetic reaction centre, where the energy is used as part of a chain of processes, including the fastest chemical reaction known, taking one trillionth of a second.

There are significant differences between animal and plant cells, for example. Although all eukaryotic cells have an outer membrane, plant cells have a cell wall on the outside of the membrane containing the structuring carbohydrate chain cellulose, which gives plants their rigidity. Because animal cells are not so rigid they can transform in shape or, in the case of cells such as white blood cells, can even engulf other material.

Unlike prokaryotes, the majority of eukaryotes reproduce sexually, allowing the mixing of genes from the parents to give variation. Although the vast majority of the genes are found in chromosomes in the nucleus of the eukaryotic cell, mitochondria also contain a few genes, thought to be the remnant of the DNA of the bacteria from which mitochondria developed. This mitochondrial DNA is unique in only originating from the female parent.

A WEALTH OF SPECIES

Although there are many prokaryote species and virus types, the diversity of eukaryotic organisms is far greater, particularly as eukaryotes are capable of forming the multicellular organisms we know as animals, plants and fungi. It has been estimated that there are around 8.7 million eukaryote species, though inevitably this figure is largely guesswork. Thanks to the diversity of insects and other invertebrates in the animal kingdom, animals dominate these estimates at around 7.8 million, with 0.6 million fungi and a mere 0.3 million plant species. It's thought that around 70 per cent of plants have been described, but only 12 per cent of animals and 7 per cent of fungi.

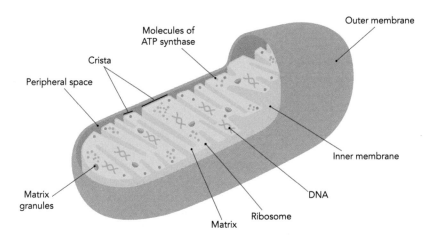

EVIDENCE

The evidence for the nature and structure of eukaryotes is identical to that for prokaryotes and viruses. The period of time since the first splitting of eukaryotes from their ancestors is primarily derived from fossils and tracking the changes in shared DNA. Zoology, natural history and botany have provided most evidence for the nature of specific eukaryotic organisms.

Opposite: Plant cells showing numbers of small green chloroplasts.
Above: Typical mitochondrion structure.

CRITICS

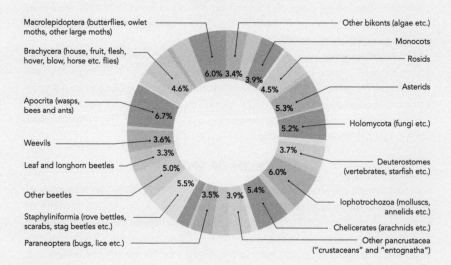

Macrolepidoptera (butterflies, owlet moths, other large moths)

Brachycera (house, fruit, flesh, hover, blow, horse etc. flies)

Apocrita (wasps, bees and ants)

Weevils

Leaf and longhorn beetles

Other beetles

Staphyliniformia (rove beetles, scarabs, stag beetles etc.)

Paraneoptera (bugs, lice etc.)

Other bikonts (algae etc.)

Monocots

Rosids

Asterids

Holomycota (fungi etc.)

Deuterostomes (vertebrates, starfish etc.)

Iophotrochozoa (molluscs, annelids etc.)

Chelicerates (arachnids etc.)

Other pancrustacea ("crustaceans" and "entognatha")

6.0% 3.4% 3.9% 4.5% 5.3% 5.2% 3.7% 6.0% 5.4% 3.9% 3.5% 5.5% 5.0% 3.3% 3.6% 6.7% 4.6%

Many of the arguments in the field have been over how to divide up the eukaryotes and over the way that they have evolved their complexity.

The division of kingdoms into animals, plants and fungi is largely non-controversial, but in comparison to, say, the standard model of particle physics, there is nothing like consensus over how the rest of the eukaryotes should be divided up. This reflects, in part, the arbitrary nature of the taxonomic hierarchy, which typically divides life into domains (archaea, bacteria and eukarya), kingdoms, phyla and so forth.

EUKARYOTIC EVOLUTION

While the evolution of eukaryotic cells from the combination of an archaeon and bacteria providing its mitochondria is now relatively uncontroversial (though there are still alternative theories), there are other aspects of the origin of the complex cell structure that are still disputed. So, for example, Lynn Margulis did not limit her theory to a single example of endosymbiosis (where one organism lives inside another to mutual benefit, eventually becoming part of a whole), but suggested other features than mitochondria, such as the movement mechanism of some single-celled eukaryotes, and the plastids that contain the chloroplasts that enable some eukaryotes to make use of photosynthesis, were originally separate prokaryotic organisms. This "serial endosymbiosis" hypothesis has not been accepted to anything like the same degree and is still regarded as unlikely by many biologists.

Above: Proportion of different eukaryote species.
Opposite: Eukaryotes encompass the main life forms we directly experience, including mammals.

WHY IT **MATTERS**

If the study of prokaryotes and viruses gave us an understanding of the earliest forms of life on Earth, plus the bacteria, archaea and viruses that can attack us or help us today, the study of eukaryotes includes all the main lifeforms we directly experience, including, of course, the human species.

Understanding eukaryotic cells at the functional level has been crucial to a better understanding of diseases and the operation of the body at the molecular level, while the whole of zoology, natural history and more is both a scientific arena entirely dependent on eukaryotes and one that provides for many people the most accessible and enjoyable aspect of science, as the popularity of TV nature documentaries attests. Speak to practically any children and their eyes will light up at the mention of dinosaurs – another example of the appeal of eukaryotes. Dinosaurs may not be cute, but they certainly appeal.

Although bacteria and viruses are particularly good at making their presence felt, it is inevitable that we regard the more complex lifeforms as both more interesting and more important, both for the future of the planet and for our own existence. In the end, given that we *are* eukaryotic, complex organisms, it is inevitable that understanding eukaryotes is a topic that will continue to be regarded as being of high importance.

FUTURE **DEVELOPMENTS**

The basics of eukaryotic lifeforms are reasonably well understood – however, many aspects of biology are still in need of further exploration, as biological systems are so much more complex than the kind of systems studied by physics.

We are still not clear how complex eukaryotic cells evolved from simpler prokaryotic cells, as they appear to have done around 2 billion years ago. Similarly, there is still plenty in the complex cellular machinery of these complicated cells that is not fully understood.

As we have seen, it is estimated that there may be around 7.8 million animals, and plenty of other eukaryotes, that are still undescribed. Although there are far fewer protists (chromista and protozoa) and many are single-celled, they remain complex organisms in comparison with prokaryotes. Especially fascinating are those that cross what we generally regard as distinct divisions. An example are the chromista, which seem to have more in common with animals than plants, in being mobile and active – yet are able to make use of sunlight to gain energy through photosynthesis.

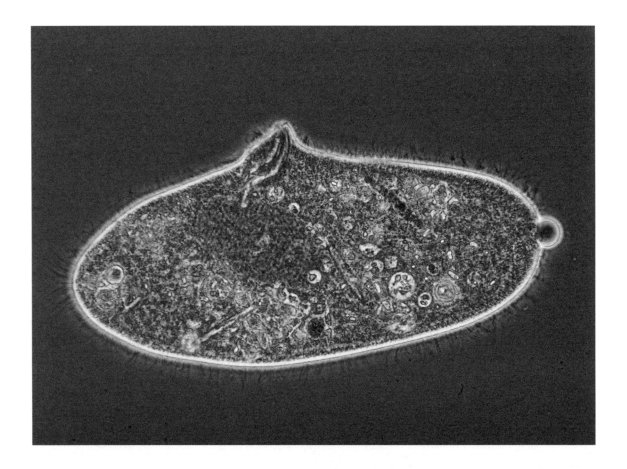

Above: Despite being single-celled, protists are still complex structures.

THE **ESSENTIAL** SUMMARY

ORIGINS	KEY THEORIES AND EVIDENCE	CRITICS	WHY IT MATTERS	FUTURE DEVELOPMENTS
As animals and plants are eukaryotes, we have **always** been aware of their existence.	Eukaryotes are defined by having a **nucleus** – an internal section of the cell with its own membrane.	The division of eukaryotes into animals, plants and fungi is non-controversial, but exactly **how other eukaryotes are divided into kingdoms is unclear**.	Studying eukaryotes tells us about **all the main lifeforms we experience directly, including humans**.	**Biological systems are so complex**, there is still much more to learn.
1666 Robert Hooke coins the term "cell" to describe the small structures in cork seen under the microscope.	Eukaryotic cells contain other smaller structures including **mitochondria**, which produce **ATP**, an energy-storing molecule.	That mitochondria were originally separate bacteria is widely accepted, but **the extension of the process to other components of the eukaryotic cell is controversial**.	Understanding the function of eukaryotic cells helps us **understand diseases** and the **operation of the body at the molecular level**.	We are **not clear how eukaryotes evolved** from prokaryotes around 2 billion years ago.
1831 Robert Brown describes components of the cell.	Some eukaryotes also contain **chloroplasts** which contain the complex mechanism for **photosynthesis**.		**Zoology and natural history are purely concerned with eukaryotes**. From dinosaurs to wildlife documentaries, they attract the public.	**New vaccines and antiviral drugs** will continue to be needed.
1860s Chromosomes first studied by a number of scientists.	A number of prokaryotes have a **propeller-like flagellum**, powered by a molecular motor.		Given that **we are eukaryotes**, they will continue to be a topic of high importance.	There are **many more eukaryotic organisms to discover**.
1890 Mitochondria discovered by Robert Altman.	Animal cells and plant cells differ, notably in that in addition to a membrane, **plants** have an outer **cell wall** reinforced with cellulose.			There is much to learn about chromista, which include **animal-like protists that are able to "eat" sunlight like plants**.
1925-48 Function of mitochondria gradually.	Eukaryotes **reproduce sexually** and have a few genes in mitochondria, originating only from the female parent.			
	There are estimated to be **8.7 million eukaryote species**, of which around 7.8 million are animals.			

DIVERSITY AND POPULATION

THE **ESSENTIAL** IDEA

"NATURE HAS PROVIDED TWO GREAT GIFTS: LIFE AND THEN THE DIVERSITY
OF LIVING THINGS, JELLYFISH AND HUMANS, WORMS AND CROCODILES."

THEODORE BULLOCK, 1996

Biological diversity and populations of different species are both constantly in a state of flux. We recognize that diversity is a good thing, which can be challenged by human intervention. When we change the environment, for example for agriculture, we tend to minimize diversity by limiting the plants that can grow and reducing animal activity.

As a result of our actions there is a concern that we could be heading into a new mass extinction. There have been a number of occasions in the past when sizeable percentages of the Earth's life were erased. The most familiar was around 65 million years ago when most dinosaurs were wiped out (the ancestors of birds survived), but there have been several others. Some believe humans now threaten a new mass extinction.

One response has been an increase in interest in rewilding, the idea of reintroducing lost species (or equivalents) to restore the balance of nature. This is, however, a difficult process as the environmental interaction of different populations was one of the earliest known examples of mathematical chaos – predicting the outcome of a change is impossible.

ORIGINS

The scientific approach to animal populations began with a book by Italian mathematician Leonardo of Pisa, better known by his nickname "Fibonacci". In *Liber Abaci* from 1202, Fibonacci introduced the "Fibonacci series": 1, 1, 2, 3, 5, 8, 13, 21… This is generated by starting with two 1s, then producing each subsequent number by adding the previous two together. Fibonacci used it to predict the population of an idealized colony of rabbits.

THE MALTHUSIAN PRINCIPLE

Leaving aside unlikely assumptions, such as rabbits never dying, the big problem with this as a picture of reality is that populations don't continue to grow forever, but are constrained by their environment, in terms of resources, competitors for those resources and predation (including disease). This was emphasized by English economist Thomas Malthus in his 1798 *An Essay on the Principle of Population*, in which he pointed out that populations grow faster than subsistence. Malthus predicted dire outcomes for humanity, though these have been avoided by advances in agricultural technology. But in the natural environment there will always be such a restraint.

In 1838, Belgian mathematician Pierre-François Verhulst produced an equation that describes the "carrying capacity of an environment", which predicts that populations should rise and plateau. But in the 1970s, Australian scientist Robert May showed when growth rate reaches a particular level, population levels become chaotic and can jump around wildly from season to season.

Although it was clear that something had happened to wipe out the dinosaurs as soon as their remains had been discovered, with the biblical flood originally suspected, the wider concept of mass extinctions only dates to the 1980s, when five were identified, the oldest being around 450 million years ago.

Although it predates the idea that we might be living through another mass extinction, the concept of rewilding, an aspect of "conservation biology", was first discussed in 1967 in a book by Canadian ecologist Robert MacArthur and American biologist E. O. Wilson. They were specifically concerned with island populations, where it was easy for native species to be devastated by human actions and imported species. By the 1990s it was becoming a more widely accepted concept, though the exact approach is still widely debated.

Opposite: The European bison has been reintroduced in a number of rewilding projects.
Top: Leonardo of Pisa (c1170–1250).
Bottom: Thomas Malthus (1766–1834).

KEY **THEORIES** AND **EVIDENCE**

NUMBERS, CHAOS AND REWILDING

"HALDANE, WRIGHT, AND FISHER ARE PIONEERS OF POPULATION GENETICS WHOSE MAIN RESEARCH EQUIPMENT WAS PAPER AND INK RATHER THAN MICROSCOPES, EXPERIMENTAL FIELDS, *DROSOPHILIA* BOTTLES, OR MOUSE CAGES."

THEODOSIUS DOBZHANSKY, 1955

In Fibonacci's simple model, population numbers grow according to the series 1, 1, 2, 3, 5, 8, 13, 21… If rabbits take one month to mature and produce a pair of rabbits each month after a one-month gestation – one rabbit of each sex – then we can see how this occurs. Starting with a single immature pair (1), after one month they mature, but there is still only one pair (1, 1). A month later they give birth to a new baby pair, so now there are 2 pairs (1, 1, 2). Next month, the first pair gives birth again and the second pair matures (1, 1, 2, 3). The following month both the first and second pair give birth (1, 1, 2, 3, 5). And so on.

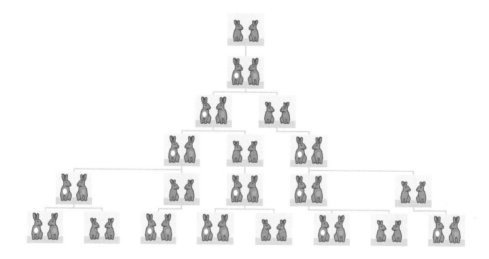

This series does crop up in nature in, for example, the layout of sunflower seeds. But it is far too simplistic for a real population. Not only would a population based on this get bigger and bigger forever, there are no deaths and there is an unlikely regularity of births. Most importantly there is no environmental influence. The real world, of course, is very different.

It seems inevitable that the population will grow relatively unchecked until food, space or predators limit it, at which point it should level off to some kind of relatively stable progression from season to season. A very simple version of this model can be represented by the equation:

$$x' = rx(1-x)$$

Here x is the current population, measured as something between 0 (extinct) and 1 (maximum carrying capacity of the environment). The factor r is the rate of growth from period to period and x' is the new population at the end of that period. Unlike the Fibonacci population, this one is constrained by the environment as a result of the $(1-x)$ part – as the population grows, there is increasing resistance to that growth.

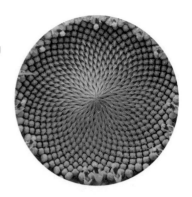

PREDICTING CHAOS

In practice, more sophisticated variants are used – but even with this simplistic model, something remarkable happens. If the growth rate, r, is increased over 3, the population level starts to jump up and down in a random-feeling fashion, demonstrating the mathematical phenomenon known as chaos.

The complexity of population growth in an environment has been combined with the knowledge of mass extinctions to produce a concern for the diversity of life on the planet. We know of five major extinction events where around 70 per cent of species became extinct. The earliest, Ordovician-Silurian, occurred around 450 million years ago, while in the latest, the Cretaceous-Paleogene around 65 million years ago, we lost most of the dinosaurs. In total, including lesser extinction events, around twenty have been discovered.

These events appear to have been caused by asteroid impact, volcanic activity and climate change. There is concern that a combination of man-made climate change and environmental damage will cause a new extinction event. Attempts are being made to restrict the impact, or to reverse it through rewilding, but given the chaotic nature of environmental systems this is a slow, experimental process as useful predictions are near impossible.

EVIDENCE

The chaotic nature of interacting populations emerges from the mathematics, but has been observed a number of times, particularly within a confined environment. The evidence for mass extinctions of the past comes from the fossil record. Whether there is a true mass extinction event due to current human interaction is less certain. We have certainly wiped out some species, such as the passenger pigeon and the dodo – and many have been in decline – but some are recovering as a result of conservation efforts and it is unlikely that we will see mass extinction levels comparable with the big five events.

Opposite: Fibonacci's rabbits producing the Fibonacci series.
Top: Sunflower seeds are often distributed according to the Fibonacci series.
Middle: The best known extinction event was the Cretaceous–Paleogene.
Bottom: Vast flocks of passenger pigeons were destroyed, leading to species extinction.

CRITICS

The importance of the impact of humans on the environment and on populations of animals and plants is widely disputed by some politicians, if not by scientists, in countries where business lobbies have a strong influence. There is often a conflict between those whose livelihood depends on farming and fishing and the ecological lobby, and similarly where industries damage the natural environment. What is clear is that it is possible to overuse a natural resource, resulting in the kind of fishing quotas that are now widely imposed.

Similarly, as rewilding often involves restoring apex predators, such as wolves and bears, farmers, and to some extent the general population, are critical of the impact this approach could have in terms of risk to livestock and even human life.

It is difficult to present a definitive scientific argument here. This is partly because of the complex and chaotic nature of environments, making it difficult to predict the outcome of a particular intervention in, for example, rewilding. And it is also due to the difficulty in pinning down the specific impact of a reduction in diversity: it clearly can have negative effects, but it is difficult to give any predictions of how those negatives will be manifested.

WHY IT **MATTERS**

From a selfish viewpoint, it is important for humanity's continued existence that our environment can support the population of Homo sapiens – which, at the time of writing, is around 7.8 billion and expected to peak around 10 to 11 billion, before starting to reduce around the end of the twenty-first century. We are sensitive to potential environmental threats, and encouraging diversity and avoiding loss of populations that we depend on can help make the survival of the human species more likely.

Not everyone extends the same enthusiasm to protecting "non-useful" species – though part of the problem here is knowing just which species will be useful to us long-term, and being sure of how changes in one species can affect others due

to the interdependency of the natural environment. However, there is a strong feeling in many cultures that the preservation of nature is a good thing in its own right, meaning that conservation is highly regarded in some countries and given considerable weight in political policy and planning decisions.

Given the impact of past mass extinctions, there can be little doubt that it is important that we understood if we are, indeed, heading into a new mass extinction due to human impact on the environment and, if necessary, should be prepared to mitigate this in the hope of maintaining biodiversity.

Opposite top: The return of some apex predators causes
concern to farmers and the general population.
Opposite bottom: The return of the beaver to European
countries has had a mixed response.
Above: Animals such as the wild boar contribute to the environmental balance.

FUTURE **DEVELOPMENTS**

The study of the interaction of populations in an environment was certainly helped in one way by the realization that population numbers could become mathematically chaotic – but just as was the case with the weather, the first system discovered to be chaotic, knowing that it is chaotic does not make it easier to make predictions.

However, weather forecasters have found a partial way around this (as will be covered in more detail in the weather systems section below) by making use of what are called "ensemble" forecasts. It will never be possible to make meaningful weather forecasts more than a few days out, but ensemble forecasts allow meteorologists to attach probabilities to different outcomes – and the same is probable to become more likely for the development of populations in an environment.

Rewilding is expected to continue to increase, while attempts to reduce human impact on biodiversity in general are likely to go hand-in-hand with measures to curtail the impact of climate change.

Above: The Danube delta rewilding area in Romania.

THE **ESSENTIAL** SUMMARY

ORIGINS	KEY THEORIES AND EVIDENCE	CRITICS	WHY IT MATTERS	FUTURE DEVELOPMENTS
1202 Leonardo of Pisa (Fibonacci) establishes his series as the result of simple, unconstrained population growth of a simplistic rabbit species.	The **Fibonacci series** 1, 1, 2, 3, 5, 8, 13, 21…, where each entry after the first two is the previous two entries added together, describes a very simple growing population with no environmental constraints.	The impact of humans on the environment is **widely disputed by some politicians** in countries with a powerful business lobby.	Selfishly, for **human existence to continue** we need an appropriate environment.	**Understanding systems are chaotic** does not make them easily predictable. However, it may be possible to use the "ensemble" forecasting techniques of meteorologists on environments.
1798 Thomas Malthus shows that populations grow faster than the means to support them, so will become constrained by their environment.	A simple **model for a real population constrained by its environment** makes the new population proportional to a growth rate, the previous population and the inverse of the closeness to capacity of the environment.	Rewilding, particularly where **apex predators** are reintroduced, **can cause concern** amongst farmers and the general public.	**Not everyone supports protecting "non-useful" species** – but it can be hard to predict which species are useful. **Others feel that nature conservation is important in its own right.**	**Rewilding** is expected to increase.
1838 Pierre-François Verhulst produces a simple equation describing the behaviour of a population in an environment with a carrying capacity.	Such a simple model **becomes chaotic, with the population jumping around wildly**, if the growth rate is over 3.	Because of the **chaotic nature of environmental systems**, it can be **hard to predict the outcome** of rewilding and other attempts to re-engineer the environment.	Given the impact of previous mass extinctions **it is important that we understand our current state** and, if necessary, mitigate any imminent extinction event.	**Attempts to minimize human impact on biodiversity** is likely to go hand-in-hand with measures to curtail the impact of climate change.
1970s Robert May shows that Verhulst's equation produces chaotic population numbers if growth is fast enough.	Five **mass extinction events** have been discovered from the fossil record, where at least 70% of species were wiped out.			
1980s Five mass extinctions in the past identified.	Extinction events were primarily **caused by asteroid impact, volcanic activity and climate change**.			
1990s Rewilding (conservation biology) is first practised.	There is concern that we are currently in a **human-induced mass extinction event**, but attempts are being made to counter it, including rewilding.			

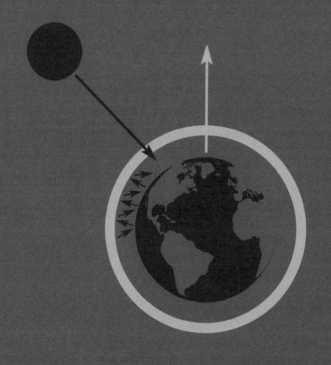

04

EARTH

●●●●●

FORMATION

THE **ESSENTIAL** IDEA

"THE FORMATION OF PLANETS IS LIKE A GIGANTIC SNOWBALL FIGHT.
THE BALLS BOUNCE OFF, BREAK APART, OR STICK TOGETHER, BUT IN
THE END THEY ARE ROLLED UP INTO ONE ENORMOUS BALL..."

CLAUDE ALLÈGRE, 1992

The Earth formed with the rest of the solar system as dust and gas were pulled together by gravity. This material was a mix of hydrogen and helium from the Big Bang with matter from exploding stars. The process would have begun around 5 billion years ago, with the basics of the solar system, dominated by the Sun and with eight major planets, in place by around 4.54 billion years ago.

At some point in those early years it is thought that a smaller planet, Theia, perhaps the size of Mars, collided with the proto-Earth. This both altered the inner structure of the young planet and blasted a significant amount of the collision debris from both planets away from the Earth's surface, where it formed the planet's unusually large moon.

Initially, much of the Earth's surface would have been molten rock, riven by volcanic activity, which pumped out much of the early atmosphere, thought to be a mix of nitrogen, carbon dioxide, water vapour and sulfur dioxide.

ORIGINS

Early ideas of the Earth's formation were based on creation myths, which typically made the Earth a flat surface with a roof-like sky, sometimes floating in water, formed by one or more deities.

As early as the fifth century BC, ancient Greek philosophers realized that the Earth was a sphere, evidenced, for example, by the way that ships at sea appeared to rise up over the horizon. Although the concept of a spherical Earth continued to be the educated view from then on (the idea that the Earth was thought in medieval times to be flat is a myth), the formation of the Earth would still largely be left in the hands of the deity.

However, by the nineteenth century, the assumed age of the Earth was starting to be pushed back. As geologists began to understand the way that layers in the Earth's surface provided a view back in time, they realized the Earth must be far older than first thought – millions of years at least. This was embarrassing, as the best estimates at the time for the age of the Sun were lower, yet the Sun was surely not younger than the Earth. For example, in 1779, the French natural philosopher Georges-Louis Leclerc, Comte de Buffon estimated the Sun's age at 75,000 years.

A SHORT TIME IN THE SUN

By 1862, Scottish physicist Lord Kelvin, after dismissing the idea that the Sun could be powered by burning coal, meaning it would last only 20,000 years, set out to make a more realistic estimate. Making use of an idea from German natural philosopher Hermann von Helmholtz that the Sun was powered by gravitational contraction, Kelvin came up with an age for the Sun of between 20 million and 60 million years. Bizarrely, though, when he calculated the implications for the Earth, based on assumptions about its original temperature and rate of cooling, he came up with 100 million years. Even this was not enough for Darwin, though, who believed that life on Earth required hundreds of millions of years to evolve to its present state.

Opposite: The Moon is thought to have been created in a collision with a smaller planet, Theia.
Top: Georges-Louis Leclerc (1707–1788)
Bottom: Mayan frieze at El Mirador depicting a creation myth.

KEY **THEORIES** AND **EVIDENCE**

DUST, WATER AND COLLISIONS

Our best model of how the solar system began was from a cloud of gas and dust, which over millions of years was pulled together gravitationally. The majority of the material (over 99.9 per cent by weight) ended up as the Sun, but a disk-shaped cloud would have remained swirling around the young Sun. As particles collided, they would have gradually gone through a process known as accretion, first into boulder-sized rocks, which would then have tended to cluster and collide to eventually form the planets.

Near the Sun, the majority of the materials would have been relatively heavy – it was too hot for many lighter materials, which would have remained as gases, and so planets would only have been able to form further out from the Sun. As a result, the Earth – the third planet from the Sun, formed around 4.54 billion years ago – is rocky. It is thought that the Earth's surface water arrived later as it would not have survived when the Earth was newly formed.

THE EARLY EARTH

The early Earth would have had a molten surface. Although iron is the most abundant element on the Earth, it is also relatively heavy, so the largest proportion of the iron would have ended up in the Earth's core. Exactly where the water came from is still disputed. For a long time it was thought to have come from comets, some of which are primarily frozen water. However, analysis of comet ice has shown that they don't have a similar mix of hydrogen and its isotope deuterium – it is now thought that only a tenth of our water came from comets.

Some of the water could have appeared during the Earth's formation from impacts which left the water sufficiently far under the surface to survive the subsequent formation of the Moon, though opposing theories suggest it was delivered after this tumultuous event, perhaps from icy asteroids which had a more Earth-like mix of hydrogen and deuterium than comets – or even that it arrived with the planet-like structure Theia, of which more in a moment.

Initially, the Earth was without a moon. However, a relatively few million years after the Earth formed, it is thought to have had a collision with another planet-sized body. The result was that a sizeable chunk of a mix of the Earth and this secondary planet, sometimes called Theia, was flung out. Consequently, the Earth has a moon that is significantly larger than might be otherwise expected for the size of the planet. It is the fifth largest in the solar system (the bigger moons all orbiting Jupiter or Saturn), and is around a quarter the size of the Earth.

Mars's moons, for example, are relatively tiny and thought to be captured asteroids. The size of the Moon has had a significant impact on the Earth: for example, in influencing the tides and helping to stabilize the Earth's orbit.

EVIDENCE

The age of the Earth is determined by radioactive isotope dating, most accurately using samarium–neodymium dating. This involves the radioactive samarium-147, which decays to neodymium-143 with a half-life of around 100 billion years, making the age of the Earth well within its scope. The pairing is chosen as such dating needs both the elements to stay together, which tends to be the case with samarium and neodymium.

The process of formation of the Earth with the rest of the solar system was determined by a mix of an analysis of the elemental composition of the Earth and the ability to compare it with rocks from the Moon and other planets, notably Mars. As for the early atmosphere, until relatively recently it was thought to have a different composition, but evidence for the suggested mix is based on the discovery of ancient zircon (zirconium silicate) crystals, which are good at trapping particles from the atmosphere as they cool after formation from volcanic magma.

Opposite top: Artist's impression of solar system formation.
Opposite bottom: Some of Earth's water may have come from icy comets.
Above: The Earth's moon is significantly larger than would normally be expected.

CRITICS

Although the scientific evidence for the Earth having formed 4.54 billion years ago is extremely strong, there is still a sizeable minority who believe that the Earth formed around 6,000 years ago, based on a seventeenth-century attempt by the Irish archbishop James Ussher to calculate the age of the Earth based on chronologies in the Bible.

The broad picture of the Earth's formation with the rest of the solar system is widely supported, but there are still a range of views on how the Earth–Moon system came into being. While the impact of Theia is the most widely accepted view, it has critics. An alternative theory was that the Moon was a separate body that was captured. However, the Earth and the Moon are extremely similar in their mix of radioactive materials – and quite different from every other planet in the solar system.

Not only does this rule out a capture scenario, it seems unlikely that Theia would have been the same in this regard as the proto-Earth. One possible way around this would be if the collision were so drastic that both early planets were reduced to rubble and mixed. Some models that simulate this happening require not one collider, but two.

Above left: James Ussher (1581–1656)
Above right: The Theia theory for the formation of the Moon is not universally accepted.
Opposite: The Earth is the only home of life we know of so far in the universe.

WHY IT **MATTERS**

Knowing how and when the Earth formed is not hugely practical, but there are three significant reasons to work this out. Firstly, it is the pre-history of our world, the only home of life we so far know of in the universe. Clearly, without the Earth having formed, there could be no life here.

Secondly, it gives us an upper limit for the existence of life on Earth, which is valuable in giving context to how and when life began, and whether it is likely to be present on other planets.

Finally, as the origin of the Earth was historically tied into creation myths, getting a better picture of the Earth's formation is also an important milestone in our ability to understand the world through science. There still tends to be a greater resistance to science, technology and modern medical practices in some areas of the world where creation myths still hold sway.

"THE ORIGIN OF THE EARTH WAS HISTORICALLY TIED INTO CREATION MYTHS"

This links with the observation that Darwin was a player in the discussion of the age of the Earth. Unless the Earth is at least hundreds of millions, and probably billions, of years old, it seems highly unlikely that evolution could produce the diversity of life that we see on Earth, let alone the range and scale of the fossil record. Darwin's understanding of evolution made it necessary for the world to be older than was widely thought at the time – and the evidence has proved him correct.

FUTURE **DEVELOPMENTS**

The age of the Earth is strongly established through radioactive isotope dating, though our models of planetary formation are still being enhanced. There seems little danger that we will deviate much from the age estimate for Earth of around 4.54 billion years. However, as we have seen, there is still considerable doubt about the mechanism by which the Earth–Moon system was brought into being.

From uranium–lead dating of Moon rocks, it seems likely that the Moon is around 4.51 billion years old, forming early in the life of the Earth, but we do not have good enough evidence to decide between the alternative formation theories.

We know a lot about the composition of the surface of the Earth and the Moon, but less about their interiors. As we will discover in the next section, the Earth's composition is surprisingly well understood, given the limitations of our ability to extract material from deep inside. However, such analysis does not enable us to distinguish potential material from an impactor such as Theia, except relatively near the planet's surface. It may be in the future that we will be able to sample from deeper inside the Earth to get a better picture.

Above: The surface of the Moon from the *Apollo 11* lunar module.

THE **ESSENTIAL** SUMMARY

ORIGINS	KEY THEORIES AND EVIDENCE	CRITICS	WHY IT MATTERS	FUTURE DEVELOPMENTS
Prehistory Creation myths describe deities producing the Earth as a flat surface with a roof-like sky. **5th century BC** Ancient Greeks discover that the Earth is spherical, but creation remains unexplained. **1779** Comte de Buffon estimates the Sun's age at 75,000 years. **1862** Lord Kelvin estimates that the Sun would only last 20,000 years if burning like coal. If its heat came from gravitational contraction, he made it 20–60 million years old, but the Earth about 100 million years. **1860s** Charles Darwin realizes the Earth must be far older for evolution to have produced current level of diversity of life.	The Earth **condensed from a cloud of dust and gas** around 4.51 billion years ago, after the formation of the Sun. The early Earth's surface was **too hot for water to exist**. It is likely to have arrived later in asteroids and other impactors. **Initially the Earth was without a moon** – our unusually large moon is thought to have formed as a result of a collision with Theia, a Mars-sized body. The dating of the **age of the Earth** comes from measuring the **decay of radioactive samarium**. **Analysis of rocks** suggests the process of Earth's formation, while zircons establish the mix of the early atmosphere.	A **sizeable minority believe that the Earth was formed 6,000 years ago** based on biblical chronology. The formation of the Earth–Moon system by impact with **Theia has critics** due to the **similarity of the materials in the Earth and Moon**. There **may have been more than one collider** in the formation of the Earth–Moon system.	Without the Earth forming as it is, **there would have been no life** (and no us). The age of the Earth gives an **upper limit for the existence of life**. Getting a better picture of the age of the Earth **moves us away from some misleading creation myths**. The age of the Earth is **important for evolution** to have had time to act.	There is uncertainty about the mechanism of the Earth–Moon system forming. **Better knowledge of the interior make-up of the Earth** may help. Current **models of planetary formation are still quite new** and could be developed further.

STRUCTURE

THE **ESSENTIAL** IDEA

"IN THE COURSE OF THE HISTORY OF THE EARTH INNUMERABLE
EVENTS HAVE OCCURRED ONE AFTER ANOTHER, CAUSING CHANGES
OF STATE, ALL WITH CERTAIN LASTING CONSEQUENCES."

CARL BERNHARD VON COTTA, 1867

The Earth is not a uniform sphere: it has four distinct layers. At its heart is a solid inner core, surrounded by a liquid outer core. Beyond this is a solid but slightly mobile region known as the mantle, with the outer crust a relatively thin layer on the outside. For that matter, it is not perfectly spherical (even allowing for the bumps and dips of mountain ranges and canyons) – because of its rotation, it is a little fatter around the equator, forming the shape of an oblate spheroid.

Working outwards from the centre, each layer of the Earth gets cooler, ranging from around 5,400°C (9,806°F) in the centre to a surface temperature of around 14°C (57°F). The Earth's inner heat comes from a combination of residual heat from the planet's formation and radioactive heating from the decay of isotopes such as uranium-238 and potassium-40.

Rather than being a single continuous layer, the outer crust is in the form of a series of regions known as tectonic plates, which gradually move.

ORIGINS

The ancient Greek idea of the Earth's structure was linked to the theory of the four elements devised by fifth-century BC philosopher Empedocles and refined by Aristotle in the following century. Each of the elements was thought to have a different tendency to move towards or away from the centre of the universe (and hence the Earth). This meant there should have been a sphere of earth, surrounded by a sphere of water, then air and then fire.

Such a model would leave all the land under water, so it was thought the earth sphere was off centre, producing a single land mass. Of course, the planet Earth was not thought to consist solely of the element earth, but rather various compounds producing the mineral structures then known.

THE EARTH'S STRUCTURE

There was not much advance until the seventeenth century. In 1692, English astronomer Edmond Halley proposed that the Earth consisted of a series of concentric spheres with gaps between them, to account for Isaac Newton's incorrectly low estimate of the Earth's density.

Similarly, it would not be until the twentieth century that most of our ideas of the Earth's structure were developed. This was because we have only experienced a tiny outer layer of the planet. The deepest we have ever drilled into the crust is the Kola Superdeep Borehole in Russia. This penetrated 12.3 kilometres (7.6 miles), less than 0.2 per cent of the way to the Earth's centre.

In 1912, German geophysicist Alfred Wegener proposed the concept of "continental drift". At the time, his idea, the basis for plate tectonics, was almost universally rejected: it would not be until the 1950s that sufficient evidence had been gathered to make his theory accepted. As for our knowledge of the inner structure of the Earth, this would initially be derived from measurements of the vibrations produced by earthquakes.

It was the detection of waves from earthquakes passing through the Earth that enabled English geophysicist Harold Jeffreys to suggest in 1926 that the Earth had a liquid core, followed in 1936 by Danish seismologist Inge Lehmann, who detected an inner solid core within that liquid region.

Opposite: The interior structure of the Earth.
Top: Alfred Wegener (1880–1930).
Middle: Seismograph trace showing waves from earthquakes.
Bottom: Edmond Halley (1656–1741).

KEY **THEORIES** AND **EVIDENCE**

LAYERS, PLATES AND TECTONICS

The Earth is made up of four roughly spherical layers (though the outer one is far less homogenous). The innermost core, around 1,200 kilometres (750 miles) in diameter, is solid, primarily consisting of iron and nickel. It remains solid despite existing at temperatures comparable to the surface of the Sun – 5,400°C (9,800°F) – because of the immense pressure, around 3 million times the atmospheric pressure we experience.

Surrounding this is the outer core, 6,800 kilometres (4,200 miles) in diameter, also primarily iron and nickel. Although the temperature is a little lower here, ranging from around 4,000°C (7,250°F) at the outer limits to that of the inner core when they come into contact, the lower pressure at this level means that the outer core is liquid. It is flowing eddies in the outer core that produce a dynamo-like effect that generates the Earth's magnetic field.

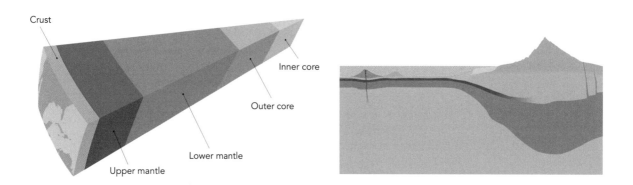

Next out is the biggest layer, the mantle, which amounts to over 80 per cent of the Earth's volume. As the outer crust is only between 0 and 100 kilometres (60 miles) thick, the diameter of the mantle is only a little below that of the whole Earth, which has a diameter of around 12,750 kilometres (7,900 miles). The temperature range of the mantle is also the greatest of the layers, ranging from the temperature of the core to around 500°C (930°F) at the outside. Strictly speaking, the rocky mantle divides into two, with a relatively thin upper mantle, which is more closely related to the crust, and the thicker lower mantle.

The outer crust is the bit we are truly familiar with, sufficiently cooled to be solid. This varies considerably in composition and thickness. The crust is typically a lot thinner under the oceans, compared with the continental crust, which is also significantly less dense with more sodium- and potassium-aluminium silicate rocks, such as granite, compared with the ocean crust's greater quantity of iron-rich minerals such as basalt.

Although the mantle is hidden from us, it has a direct influence in the movement of tectonic plates. The mantle is technically solid, but is capable of flowing extremely slowly, particularly in the lower pressures of the outer mantle.

Convection currents, driven by differences in temperature, mean that the plates of the surface are moved very slowly.

The idea that such movement was possible was first hypothesized because of similarities between coastlines on different continents, suggesting that they were once joined. There are around seven major plates and many smaller ones, which move against each other at rates no faster than 0.1 centimetres (0.04 inches) a year. Where plates are moving towards each other laterally, one can flow under the other (subduction), forming mountain ranges and high risk of earthquakes. Where plates move away from each other, ocean basins open and volcanic activity can be intense. There are also plates that move against each other, producing faults liable to earthquakes, such as the San Andreas Fault in California.

EVIDENCE

Although earthquakes can be a cause of terrible destruction, they have also been the most important tool in developing an understanding of the inner structure of the Earth. An earthquake sends a series of powerful waves right through the planet, and by studying the way these change form and direction it has been possible to deduce a lot. Earthquakes produce two different types of waves in the body of the Earth. The existence of a liquid core, for example, was deduced from the way that longitudinal (squashing in direction of travel) P-waves were refracted by it, but transverse (moving side to side in the direction of travel) S-waves were stopped.

Evidence for tectonic plates came from palaeomagnetism. This uses iron-bearing rocks, which store traces of magnetism from the Earth's magnetic field. Measurements of rocks in different locations suggested that the land they were on had moved with respect to the Earth's poles. Combined with more direct evidence of movement – for example, in parts of the ocean crust – this supported the existence of plate tectonics.

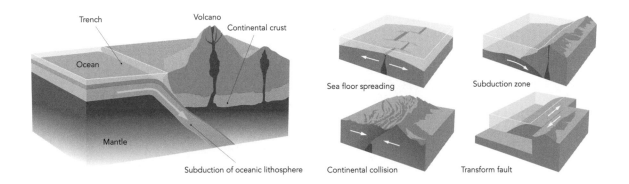

Opposite left: The four main layers with the core split into inner and outer.
Opposite right: Section through the Earth's crust.
Top: Disruption around the San Andreas fault.
Above left: A subduction zone.
Above right: The four main types of plate interaction.

CRITICS

Although the structure of the Earth's crust was subject to some debate in the seventeenth and eighteenth centuries, between those who thought that the layers that were observed in exposed rocks and mines were the result of the biblical flood and those who thought they were due to heat from within the Earth, there was little in the way of theory of the interior structure to stimulate criticism (apart from the occasional venture into hollow Earth theories) before the current ideas were developed.

By far the biggest counterargument came as a result of the idea of continental drift. Although there is a clear similarity between the coastlines of various continents that suggested they once fitted together like jigsaw pieces, Wegener faced intense criticism when he first proposed the theory – because the continents seemed such fixed bodies. Geologists could not imagine a mechanism for the continents to move, especially as it was known that the continental crust was largely less dense than the oceanic crust.

Above: The layers in exposed rocks were thought to be the result of the biblical flood or internal heat.
Opposite top: Earthquake damage on Haiti.
Opposite bottom: The protective effect of the Earth's magnetic field.

WHY IT **MATTERS**

Aside from the desire to understand how our planet works, there are two major impacts from the Earth's structure that demand our attention.

The first is the seismological influence. Volcanoes and earthquakes are amongst the most devastating of the causes of natural disasters. Both are driven by structural activity within the planet, particularly from the impact of plate tectonics. It is also important in understanding volcanoes to have information on the temperature gradients within the Earth and the materials beneath the surface.

The second reaches much further into the Earth to the liquid core. The magnetic field that is generated there is hugely important for us. The most obvious application may be our ability to use compasses, but the Earth's magnetic field has a much more important role in keeping us safe from space radiation.

The Sun blasts towards us a stream of high-energy electrically charged particles known as the solar wind. The Earth's magnetic field deflects the solar wind, protecting our atmosphere. Were the magnetic field not there, we would expect that the solar wind would strip off the ozone layer, exposing us to life-threatening levels of ultraviolet radiation and, over time, could strip away the entire atmosphere. It is thought that Mars once had a significant atmosphere, but lost it as a result of its magnetic field failing.

FUTURE **DEVELOPMENTS**

As we have seen, the further we go into the Earth, the higher the temperature. A temperature gradient like this is a source of energy – heat engines work by making use of a difference in temperature between two locations. This gives the opportunity of a different kind of green energy source which, unlike all the others (except nuclear), does not rely on energy from the Sun. This is geothermal energy.

"GEOTHERMAL ENERGY IS PARTICULARLY EASY TO MAKE USE OF WHERE THERE ARE EXISTING OPENINGS IN THE CRUST, SUCH AS VOLCANIC ACTIVITY."

Simply by drilling down into the Earth it is possible to reach higher and higher temperatures. Geothermal energy is particularly easy to make use of where there are existing openings in the crust, such as volcanic activity. As a result, Iceland is the world leader in the use of geothermal energy, but it could be deployed significantly more in other countries that are limited in their ability to use solar energy.

In terms of our basic knowledge, we are still finding things out about the Earth's inner structure. Bear in mind that practically nothing was known before the twentieth century. As mechanisms for interpreting waves passing through the Earth get better, so will our understanding of the details of the Earth's structure.

Above: Hellisheidi geothermal energy plant in Iceland.

THE **ESSENTIAL** SUMMARY

ORIGINS	KEY THEORIES AND EVIDENCE	CRITICS	WHY IT MATTERS	FUTURE DEVELOPMENTS
4th century BC Ancient Greeks thought the Earth had concentric spheres of earth, water, air and fire.	The Earth consists of **four roughly spherical layers**.	The source of the **structure of the crust** was debated between being caused by the biblical flood or heat from within the Earth.	It is **understanding how our planet works**.	The heat differential between the surface and deeper into the Earth's structure can be used as a source of **geothermal energy**.
1692 Edmond Halley proposed the Earth was made of concentric hollow spheres to account for its (incorrect) density.	The **inner core**, 1,200 km in diameter, is **solid iron and nickel** at around 5,400°C.	The biggest criticism was of **continental drift**, as **no mechanism to explain it** could originally be envisaged.	We need to know about the Earth's structure to **understand the source of volcanoes and earthquakes**.	We are still relatively **early in our study of the interior structure** and can expect more to emerge as our probes and wave detection get better.
1912 Alfred Wegener suggests that the continents could move in "continental drift".	The **outer core**, 6,800 km in diameter, is **liquid iron and nickel** and **produces the Earth's magnetic field**.		The Earth's magnetic field, originating in the outer core, protects us from the **solar wind** which could otherwise **strip away the ozone layer and eventually the atmosphere**.	
1926 Based on seismology data, Harold Jeffreys suggests the Earth has a liquid core.	The **mantle** is over 80% of the Earth's volume. Though **solid rock, it is capable of very slow movement** due to convection currents.			
1936 Inge Lehmann discovers an inner solid core.	The **outer crust** is solid. It's a lot **thinner under the oceans**, but denser there.			
1950s Evidence supports the existence of tectonic plates as the basis for continental drift.	Continental drift occurs because **tectonic plates** – large areas of the crust – **move on the mantle**, forming mountain ranges and causing earthquakes.			
	Evidence comes from **measurement of shock waves** through the Earth from earthquakes and **paleomagnetism**.			

ORIGINS OF LIFE

THE **ESSENTIAL** IDEA

"LIFE CAN BE THOUGHT OF AS WATER KEPT AT THE RIGHT TEMPERATURE IN THE RIGHT ATMOSPHERE IN THE RIGHT LIGHT FOR A LONG ENOUGH PERIOD OF TIME."

NORMAN BERRILL, 1958

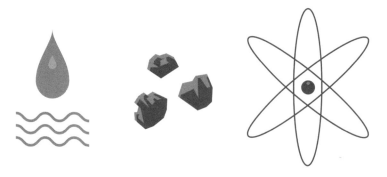

Although we have looked at life as a phenomenon within biology, life started on Earth and from the resources on the planet – so, when we consider the very beginnings of life, known as abiogenesis, it is appropriate that it appears in the Earth section. Remarkably, it seems likely that life has existed on Earth for around 4 billion years, appearing only 500 million years after the Earth formed.

All life as we know it is dependent on three key components – water, carbon and energy. Of course, there are significantly more essential elements to enable the processes of life to function, but without the unique flexibility of the carbon atom and the medium of water, life is unlikely to exist. Similarly, life would be impossible without a constant source of energy.

A number of the basic molecules required for life have been discovered out in space. Whether or not this implies that life is common in the universe depends on how difficult it was for life to start in the first place. It only appears to have happened successfully once on Earth.

Above: The essentials for life: water, carbon and energy.
Opposite top: Joseph Hooker (1817–1911).
Opposite bottom: Disproving spontaneous generation.

ORIGINS

The how and when of life starting on Earth was for a long time driven by religious beliefs, with many in the West believing that the Earth was created around 6,000 years ago, with life present from the beginning. There was, however, a parallel strand dating back to Aristotle, who in the fourth century BC suggested that lesser organisms could be produced by spontaneous generation – appearing from what were usually unpleasant surroundings.

There was a degree of logic to this theory. Leave a piece of rotting meat for a while and maggots emerge from it. A classic experiment disproving spontaneous generation was undertaken in 1668 by Italian physician Francesco Redi, using containers with the same rotting meat, some open, some covered in cloth – only the open containers generated maggots, suggesting that flies were laying unobserved eggs on the meat. It was not until the nineteenth century that spontaneous generation was entirely dismissed.

GEOLOGICAL TIMESCALES

The developing understanding of geological timescales in the nineteenth century, combining with Charles Darwin's requirement of many millions of years for evolution, pushed back the date for the origins of life to a more distant past. But this didn't provide a mechanism for life coming into existence if divine intervention was set aside and spontaneous generation proved unlikely.

Darwin wrote in 1871 to his friend Joseph Hooker, "It is often said that all the conditions for the first production of a living being are now present, which could ever have been present. But if (and oh what a big if) we could conceive in some warm little pond with all sorts of ammonia and phosphoric salts, – light, heat, electricity present, that a protein compound was chemically formed, ready to undergo still more complex changes…"

In his letter, Darwin seemed to propose a special case of spontaneous generation in conditions different from our current environment, a theory later developed by a number of scientists, including Alexander Oparin in Russia in 1924 and John Haldane in England in 1929, suggesting that a primordial "soup" (Haldane's term) resulted in chemical reactions producing the chemicals necessary for life.

This concept was reinforced in 1952, when American chemist Stanley Miller experimented with sending electricity through a mix representing the atmospheric conditions on the early Earth and water. The result was the production of a number of amino acids – chemical compounds necessary for life.

Flies entered and laid eggs that hatched maggots

No flies entered, but they laid eggs on the gauze that hatched maggots, or eggs fell through the gauze and hatched on the meat

No flies, maggots or eggs could enter

KEY **THEORIES** AND **EVIDENCE**

CARBON, WATER AND ZIRCONS

"LIFE EXISTS IN THE UNIVERSE ONLY BECAUSE THE CARBON ATOM POSSESSES CERTAIN EXCEPTIONAL PROPERTIES."

JAMES JEANS, 1930

It is thought that life emerged around 4 billion years ago. That it was able to do so is highly dependent on the chemical properties of carbon. As we have seen in the chemistry section, carbon is by far the most flexible of the elements in the way that it bonds to produce different structures, ranging from the simplicity of methane up to the complexity of DNA and the proteins that are so essential for life. It is generally agreed that without carbon and a means to accumulate carbon into appropriate structures, there could be no life.

Luckily, carbon is not an uncommon element, and basic organic structures form easily, as is evidenced by the discovery of a number of simple organic molecules and even some amino acids in space. Another relatively common requirement, one that we are particularly well endowed with on Earth, is water. It's not for nothing that Earth is a blue planet from space.

Water has two vital features that make it so important to the existence of life. One is that it is a good solvent. This is because the water molecule is "polar" – both hydrogen atoms are on the same side of the molecule, so a water molecule has a relative positive electrical charge on one side and a relative negative charge on the other side. We have already seen how this makes hydrogen bonding between water molecules possible, but it also means that water is good at latching on to other atoms and molecules, the requirement to be a good solvent.

The reason this is so essential is that the processes of life require various ions and molecules to be moved around and to interact with molecular machinery. Without a solvent to transport them, these parts will tend to stick together and not function.

Water's other valuable feature is that it is a liquid at a wide range of the temperatures that occur on Earth. Of course, temperatures do regularly fall below the freezing point of water, but organisms usually have mechanisms to prevent this causing their cells to solidify and cease functioning. While there seems to be no alternative to carbon for forming the structural molecules of life, in principle water could be replaced by another solvent. Some have suggested, for example, that there could be life on Saturn's moon Titan, which has clouds and surface liquids that are hydrocarbons, such as methane and ethane, which could act as solvents.

THE ORIGINS OF LIFE

While there is no single idea on how life was started, one convincing suggestion is that it first occurred in undersea vents. These are sources of hot water, rich in minerals. One of the biggest problems with life starting is how many different things have to come together at the same time. Narrow vents could provide a number of the requirements – a structure to keep the life together before a cell wall developed, a mechanism to gain nutrition and excrete, a source of energy and more. There is still a significant mountain to climb for life to begin, but by removing some of the hurdles, such an environment would make an excellent starting point.

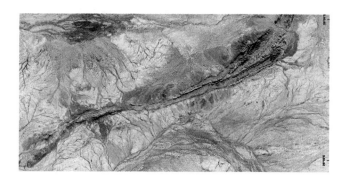 

EVIDENCE

It is suspected that life emerged around 4 billion years ago because zircons, crystals of zirconium silicate, amongst the oldest known substances on Earth, have structures that tend to trap other atoms. One such atom is carbon, and in zircons found in Jack Hills in Australia, dating back 4.1 billion years, there is an imbalance of carbon isotopes, typically produced by biological processes.

Zircons also provide evidence for the structure of the early atmosphere from the presence of compounds influenced by the atmospheric content. Although there is no direct evidence for ideas of how life was able to begin, the current best hypotheses provide logical suggestions for mechanisms to reduce the barriers to life being able to spontaneously begin.

Opposite left: The shape of the water molecule makes it polar.
Opposite middle: The complex structure of a protein.
Opposite right: Artist's impression of a hydrocarbon lake on Titan.
Above left: Jack Hills, Australia.
Above right: An undersea vent pumping out hot water.

CRITICS

The criticism of a natural origin for life was summed up in the work of English clergyman William Paley. In his 1802 book *Natural Theology*, Paley made the watchmaker argument. He pointed out that if he came across a stone while crossing a heath then it might have been there forever. But if he found a watch, he would not think the same. This was because the watch was put together for a purpose and its parts had to be as they were to fulfil its function.

The argument, then, was that a complex structure like a living organism would not spontaneously occur, but had to be created, just as was the case with the watch. Paley was writing when ideas of evolution were in the air, but well before Darwin set the concept out in detail. His argument applied to even the simplest form of life – there is a considerable leap from a collection of chemicals to a living organism, which still requires complex mechanisms to develop.

Miller's experiment has been effectively criticized on two levels. Since the experiment, it has been discovered that his attempt at a primeval atmosphere had the wrong mix of gases, and the experiment would have been less successful with the correct gases. But also, simply making amino acids and saying you have created life would be like making nuts and bolts and saying you have built a car. There is a big leap from this experiment to a living organism.

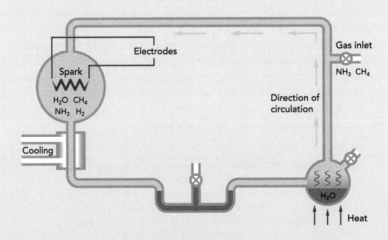

WHY IT **MATTERS**

How life began is one of the big questions that science is yet to definitively answer. Clearly this is of importance to us all, not in some attempt to recreate the moment, but rather to understand how we and the whole panoply of life around us came into being.

Arguably, the most important benefit of getting a better feel for how life began on Earth is to better assess how common life is likely to be in the universe. Opinions amongst scientists on this range all the way from thinking that most planets with appropriate conditions will have some form of life through to the hypothesis that life is extremely unusual and that we may be the only intelligent lifeforms in our galaxy.

The frequency with which alien lifeforms are encountered in science fiction emphasizes our natural fascination with the topic. We now know that there are many stars with planets, yet we have no evidence as yet that any of them harbour life. In some ways this is not surprising. The distances between stars are vast. Travelling the fastest a human ever has – 40,000 kilometres per hour (24,850 miles per hour) on *Apollo 10* – it would take over 11,000 years to reach the nearest star apart from the Sun. And for all we know, the science fiction dreams of faster-than-light space travel will always be fiction. But it would make attempts to estimate the chances of life beginning on different worlds easier if we had a better idea of how it began here.

Opposite top: William Paley (1743–1805).
Opposite bottom: The Miller experiment.
Above: The fastest humans have travelled was on *Apollo* missions.

FUTURE **DEVELOPMENTS**

Oddly, one of the most important potential developments for understanding the origin of life on Earth would be if we could find life elsewhere in the solar system. Although the stars are beyond our reach, the solar system isn't. If we can find evidence of life on other planets or moons, studying it could help provide evidence of how life began here.

Such evidence does not require panspermia. This is the theory first proposed scientifically in the nineteenth century and revived and enhanced by twentieth-century English astrophysicist Fred Hoyle and the Sri-Lankan born British astronomer Chandra Wickramasinghe. The idea is that, rather than originating on Earth, life arrived from space.

This is not a ridiculous concept. Material does arrive from space, including meteorites that originated on Mars, and in principle basic lifeforms could survive travel through space. In panspermia were true, it would do away with the need to explain the origin of life on Earth – but only pushes back the problem to another location.

However, the most likely discovery would be parallel life developing on another world. As we only have one example on Earth, this might provide significant clues as to how life began here.

Above: The surface of Mars – as yet no life has been found here.

THE **ESSENTIAL** SUMMARY

ORIGINS	KEY THEORIES AND EVIDENCE	CRITICS	WHY IT MATTERS	FUTURE DEVELOPMENTS
Antiquity onwards Much of the explanation for the origins of life on Earth involves creation by a deity.	**Water, life and energy** have proved essential for life on Earth.	William Paley's **watchmaker argument** was a powerful suggestion that it would be hard for life to start spontaneously.	Broad importance to **know where life came from**.	Though the stars are beyond our reach, **the solar system isn't**.
4th century BC Aristotle suggests that some lesser forms of life undergo spontaneous generation – usually from decaying matter.	Life's emergence around 4 billion years ago was highly dependent on **carbon's flexible chemistry**.	**Miller's experiment has been criticized** both for using the **wrong mix of gases** for what is now believed to be the early atmosphere, and because having amino acids is only as close to life as having a **box of nuts and bolts** is to having a working car.	We can better assess **how common life may be in the universe**.	Discovering **life elsewhere in the solar system** could help understand how life formed here.
1668 Francesco Redi shows that the apparent spontaneous generation of maggots on rotting meat is due to flies laying their eggs.	**Water provides a versatile solvent**, essential to carry around the many compounds required by life.		We are **unlikely to encounter aliens directly** – the distances in interstellar travel are too immense.	This does not require panspermia – **life on Earth arriving from space** – though this theory is not totally ludicrous.
1871 Charles Darwin speculates about life starting in a mineral-rich "little warm pond".	**Water is a liquid at most surface temperatures** on Earth, essential to fulfil its roles.		Knowing more about how life emerged and how easy that was would make it easier to decide **if life was likely to be common or rare in the universe**.	**Parallels and differences with life from a different source** would help us understand how life started here.
1924 Alexander Oparin suggests that a primordial soup of organic chemicals resulted in the reactions that brought about life.	Although there is no good alternative to carbon, **other solvents, such as liquid hydrocarbons, could substitute for water** on other worlds.			
1952 Stanley Miller's experiment, trying to duplicate the conditions of early Earth, generates amino acids.	**It is possible life began in undersea vents**, which provide a framework, warmth, chemicals and a flow of materials that might have lowered the barrier to life beginning.			
	Ancient zircons, trapping atoms that respond to the atmosphere and carbon levels, **give us evidence for the early start of life**.			

CARBON CYCLE

THE **ESSENTIAL** IDEA

"THE AGRICULTURALIST HOLDS IN HIS HAND THE KEY TO THE MONEY
CHEST OF THE RICH, AND THE SAVINGS-BOX OF THE POOR; FOR
POLITICAL EVENTS HAVE NOT THE SLIGHTEST INFLUENCE ON THE
NATURAL LAW, WHICH FORCES MAN TO TAKE INTO HIS SYSTEM,
DAILY, A CERTAIN NUMBER OF OUNCES OF CARBON AND NITROGEN."

JUSTUS VON LIEBIG, 1851

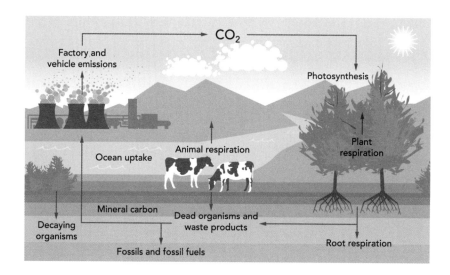

Chemical processes do not create or destroy atoms. As a result, every atom in
your body was previously in other organisms – and apart from a small ingress of
atoms from meteorites and other colliders, every atom on the Earth has been
here since the planet formed.

Since life is crucially dependent on carbon, an effective way to understand the
Earth's biological processes is to trace the cycle of carbon atoms. Carbon from the
air, in the form of carbon dioxide, is extracted by plants to provide them with the
backbone raw material for growth. Plants are eaten by animals (some of which are
eaten by other animals), transferring carbon to the animal organism. And carbon

dioxide is emitted by animal respiration, by geological emissions and by decaying plants and animals back into the atmosphere. It's a real circle of life.

Of course, there are subtleties to these processes. For example, bacteria play a significant role in the release of carbon back into the atmosphere. But the full carbon cycle is arguably the single most important process of interaction between living things and the planet.

ORIGINS

Central to understanding the carbon cycle was the realization that air was a mix of gases, including carbon dioxide. Between 1774 and 1786, English natural philosopher Joseph Priestley ran a series of experiments on what he described as "different airs". Priestley discovered oxygen, which he called "dephlogisticated air". This odd terminology was because Priestley supported a theory based on a substance that was a kind of anti-oxygen – when something burned it was supposed to give off "phlogiston".

Priestley found that a mouse "damaged" air, producing phlogiston (removing oxygen), but if he added a plant to the jar the air was restored by removing phlogiston (adding oxygen). A few years earlier, Priestley had discovered that bubbling "fixed air" (carbon dioxide) through water produced sparkling water. He did not have the commercial inclination to do much with this, leaving it to the Swiss amateur scientist Johann Schweppe to start selling it. But he did discuss his findings on various gases with Dutch chemist Jan Ingenhousz, who in 1779 observed photosynthesis in action, noticing that, when in the light, plants gave off bubbles of oxygen from their leaves.

GIVING OXYGEN ITS NAME

French chemist Antoine Lavoisier, who moved us away from the clumsily named airs, disposing of phlogiston and naming oxygen, took the basics of the carbon cycle a little further, particularly with the animal part of the cycle. For example, in 1783 he measured the carbon dioxide and heat given off by a guinea pig, compared this with the heat given off by burning carbon and deduced that respiration was a kind of slow combustion, taking in oxygen from the air and giving off carbon dioxide. By 1804, Swiss chemist Nicolas de Saussure was able to demonstrate that the carbon in plants was specifically taken from carbon dioxide in the air.

By 1847, the French chemist Jacques-Joseph Ébelmen had pieced together another significant component of the carbon cycle where volcanoes pump out carbon dioxide produced from buried organic remains, making him the first to really bring together the organic and geological components of the carbon cycle.

Opposite: The carbon cycle.
Top: Joseph Priestley (1733–1804).
Bottom: Antoine Lavoisier (1743–1794).

KEY **THEORIES** AND **EVIDENCE**

CARBON DIOXIDE, PLANTS AND ANIMALS

"YOU WILL DIE BUT THE CARBON WILL NOT; ITS CAREER DOES NOT END WITH YOU. IT WILL RETURN TO THE SOIL, AND THERE A PLANT MAY TAKE IT UP AGAIN IN TIME, SENDING IT ONCE MORE ON A CYCLE OF PLANT AND ANIMAL LIFE."

JACOB BRONOWSKI, 1965

The carbon cycle runs at two paces. The slow cycle is geological. Carbon dioxide from the atmosphere reacts with water, forming the weak carbonic acid. This can slowly dissolve rocks, resulting in water running into the sea that contains metal ions such as calcium and magnesium, plus carbonate ions (each a carbon atom with three oxygen atoms attached). Over time, these deposit out as minerals such as calcium carbonate (limestone), particularly when the ions are used by organisms that build shells, which then fall to the ocean floor.

This carbon-containing material, combined with carbon that has been deposited by dead organisms and over time has been compressed into oil, gas or coal, is eventually returned to the atmosphere by the heat and pressure of volcanic action. This whole slow cycle can take hundreds of millions of years to complete.

The more familiar life-related cycle is far quicker, dependent as it is on the life and death of organisms. Plants (and plankton in the sea) use solar energy in photosynthesis to crack open the carbon dioxide molecule and make use of the carbon in growth, disposing of oxygen into the air as a waste product. (The carbon cycle has an inverse: the oxygen cycle.)

From plants and plankton, carbon can be returned to the atmosphere in a number of ways through variants of the reaction known as combustion, where oxygen combines with carbon-containing molecules, breaking bonds and releasing both energy and carbon. The two "classic" routes are when an animal eats the plant, releasing carbon dioxide through respiration, and when a plant dies and decays, often aided by bacteria. However, carbon dioxide is also

released by plants at night, making direct use of that energy, and also when fires consume living (or once-living) matter. This last example not only involves natural combustion, but, for example, human use of fossil fuels, which were all once-living matter.

CLIMATE CHANGE

Climate change, discussed below, caused by human activity is primarily a result of the release of greenhouse gases into the atmosphere, notably carbon dioxide. From the point of view of the carbon cycle, in principle this can have the benefit of encouraging faster plant growth, but on the slow-cycle side, the result is more acidification of the oceans from dissolved carbon dioxide, which can cause serious damage to coral reefs and shell-forming sea life.

Although the carbon cycle is the most critical one for the planet, living organisms also depend on a water cycle and a nitrogen cycle. In the water cycle, water – primarily from oceans but also from fresh water sources – evaporates to become water vapour in the atmosphere. This is moved by wind to locations where it is cool enough for the water to condense and fall as rain. For land-based life, the water cycle is essential to obtain fresh water.

The nitrogen cycle reflects another major requirement of plants to grow effectively. Nitrogen from the atmosphere (around 78 per cent of its make-up) is "fixed" in the soil, typically by bacteria, some of which live symbiotically on plant roots, converting nitrogen to nitrates. After plants have made use of the nitrates in growing, just as with carbon, the nitrogen is returned to the air after the plant dies, typically through bacterial action.

EVIDENCE

As we have seen, experiments have been run since the eighteenth century noting the gas output and consumption of living things. The bigger picture emerged with an enhanced understanding of geology and volcanology for the slow cycle, and since we have been able to study bacteria to understand their part in the cycles of the Earth.

Opposite: Durdle Door, Dorset dramatically demonstrating the effects of erosion.
Above: Bioluminescent plankton in the Maldives.

CRITICS

Early criticism came from confusion caused by the phlogiston theory. Although it proved possible for scientists such as Priestley to develop some of the ideas that would form the carbon cycle without an understanding of the nature of oxygen, it was only with Lavoisier's discoveries that it was possible to develop the carbon cycle picture properly.

Some of the specifics of the carbon cycle produced considerable debate. For example, though it may seem obvious now, the idea that coal was carbon from decayed organisms was disputed when it was first put forward because without a clear picture of the timescale of life on Earth, some believed that coal was a geological product, predating the existence of life.

It is not surprising that the full picture of the carbon cycle was not available until relatively late in the nineteenth century as it is dependent on a modern atomic view of chemistry – bear in mind that, despite John Dalton's work and that of his successors, the existence of atoms would continue to be disputed throughout the nineteenth century and there was only a consensus on the matter at the start of the twentieth century.

Similarly, once the carbon cycle was recognized, there was some dispute between those supporting the slow and fast cycles, before it was realized that both made a significant contribution.

Above: Fossilized ferns in coal.
Opposite top: Understanding the carbon cycle is crucial to sustaining the environment.
Opposite bottom: Volcanic carbon dioxide output is dwarfed by human production.

WHY IT **MATTERS**

The carbon cycle is the driving force of life. Without it, we could not exist – and it is only by understanding it that we can fully comprehend the complex web of interaction that is involved in life on Earth. Although life is arguably only a tiny surface activity as far as the planet as a whole is concerned, it is by far the most complex aspect of the Earth: the carbon cycle tracks the effectiveness of the Earth's living environment.

It's easy for us to regard the world's natural resources as simply materials for us to make use of as we will, but an understanding of the carbon cycle makes it clear why, for example, plants are far more than foodstuffs and raw materials for construction or burning, but rather a fundamental part of this constant cycle of planetary renewal.

At a time when climate change is a major concern, understanding the climate cycle is crucial, as direct human production of greenhouse gases has to be seen in the wider context of both the slow and fast carbon cycles. For example, some of those attempting to downplay the significance of the human contribution to greenhouse gases claim that the slow carbon cycle puts far more carbon dioxide into the atmosphere, meaning that the human contribution is negligible – but studies make it clear that normal volcanic output is dwarfed by human production.

FUTURE **DEVELOPMENTS**

The basics of the carbon cycle are well understood, but this is such a complex system that, as is the case with other environmental factors such as populations, the fine detail is often not well known, and the impact of changes can be impossible to forecast definitively. A significant amount of the development in this area is likely to be in understanding the impact of, for example, bacterial contributions better.

As will be discussed in more detail in the climate change section, an important aspect of dealing with climate change will be "rebalancing" the carbon cycle to ensure that there is not more carbon going into the atmosphere than is taken out of it. As well as reducing emissions this can be done both by using natural means to remove carbon – for example, tree planting – and by technological means, for example using mechanisms to extract carbon from the atmosphere and lock it away underground or underwater. These approaches both need further refinement as, for example, tree planting is a very slow way of sequestering carbon, while the technology for carbon capture and storage is still at the very early stages of development.

Above: Tree planting is one of a number of ways to reduce atmospheric carbon.

THE **ESSENTIAL** SUMMARY

ORIGINS	KEY THEORIES AND EVIDENCE	CRITICS	WHY IT MATTERS	FUTURE DEVELOPMENTS
1770s Joseph Priestley experiments with "different airs" finding a mouse "damages" air, but a plant restores it. **1779** Jan Ingenhousz observes photosynthesis as oxygen bubbles off plant leaves. **1783** Antoine Lavoisier establishes that respiration is a slow combustion process taking in oxygen and giving off carbon dioxide. **1804** Nicolas de Saussure shows that the carbon in plants comes from the air. **1847** Jacques-Joseph Ébelmen brings together the organic and geological components of the carbon cycle.	There are **slow and fast** carbon cycles. The **slow, geological cycle** involves carbon dioxide from air dissolving in water, reacting with rocks and producing carbon-bearing rock which is returned to the atmosphere (along with some ancient biological carbon) by volcanoes. The **fast, biological cycle** involves plants taking carbon from the atmosphere. Plants are eaten by animals, producing carbon dioxide, as do plants and animals when they decay after death. **Man-made carbon dioxide** contributes to the cycle, encouraging plant growth but also increasing ocean acidification. Although the carbon cycle is **essential for life**, it exists in parallel with both **water and nitrogen cycles**.	Some early criticism arose from confusion caused by the pre-oxygen **phlogiston theory**. The idea that **coal was carbon from decayed organisms** was disputed as some thought coal predated the existence of life on Earth. The **carbon cycle can only really be understood with an atomic viewpoint** – which was the subject of significant criticism throughout the nineteenth century.	The carbon cycle is the **driving force of life**. Although life is only a **tiny surface activity** of Earth it is by far its **most complex system**. It's easy to think of the world's natural resources as **materials to use at will**, but we have to bear in mind their **role in the carbon cycle**. It is essential to understand the carbon cycle to **put human production of greenhouse gases contributing to climate change into context**.	The basics are well understood, but the **fine detail, such as bacterial contributions**, need better understanding. An important aspect of dealing with climate change will be **rebalancing the carbon cycle**. Mechanisms from **tree planting** to **carbon capture and storage** need further refinement.

ROCK CYCLE

THE **ESSENTIAL** IDEA

"THE ROCKS USUALLY HIDE THEIR STORY IN THE MOST DIFFICULT AND INACCESSIBLE PLACES."

ROY CHAPMAN ANDREWS, 1926

Geology, literally "earth discourse", the study of the physical nature of the Earth's composition, has its own cycle, relating the three major types of rock: igneous, metamorphic and sedimentary. Like the carbon cycle, the rock cycle involves a repeated process as rocks transition from one form to another, though the timescales are far greater and the atmosphere plays a far smaller role.

There are multiple branches available in the rock cycle, as types of rock transition from one to another. Rocks that are pushed down into the mantle by moving tectonic plates, reaching a "subduction zone" where one plate slides down underneath another, can melt into magma. When the molten rock is pushed back up, typically through a volcano, it crystalizes in the igneous form.

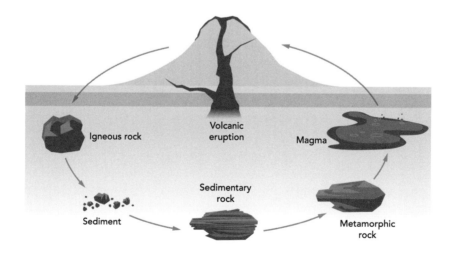

Above: The rock cycle.

Sedimentary rocks, as the name suggests, are the result of the sedimentary build-up of grains (such as sand), worn away from existing rocks by the impact of the weather. Over time and under pressure, the grains are combined into sedimentary rocks. By contrast, metamorphic rocks are those that undergo a change of crystal structure under temperature and pressure without ever becoming liquid.

ORIGINS

Two key Scottish figures lie behind the development of the rock cycle: James Hutton and Charles Lyell. Earlier there had been considerable dispute between "Neptunists" and "Plutonists". The Neptunists, notably German geologist Abraham Werner, believed that the Earth's rock layers were primarily formed by the action of a large amount of water. Plutonists, on the other hand, taking their lead from the Italian geologist Anton Moro, thought that the layers of rock were set in place by heat and fire.

CONSTANT FLUX

Geologist James Hutton promoted the idea that the Earth was in a constant (if very slow) state of flux – over many millions of years a combination of heating until molten and cooling, along with erosion and the depositing of sediments, had produced the varied and layered state of the rocks that we discover in the countryside. His ideas formed a central part of the concept of uniformitarianism. This is a misleading-sounding name, as it suggests everything is the same – but was intended to purvey a gradual cycle of change over vast timescales, rather than sudden and catastrophic change due to a single distinctive event in the past, such as the biblical flood. It's the *processes* of change that are uniform here, not the state of the rocks.

Hutton also noted that what had been seabed could now form part of the dry land, as a result of long-term cycles of deposits of material, volcanic activity and erosion. The key message of uniformitarianism was that the same processes were in action now as had caused the various layers and types of rock to emerge. Hutton's work was built on by the younger geologist Charles Lyell. Like Hutton, Lyell was a supporter of uniformitarianism and the idea that geology was the product of "deep time" stretching back eons. Lyell also studied earthquakes and volcanoes – although these often produce locally catastrophic events, he saw them nevertheless as part of the longer-term cycles that repeatedly transformed the rocks in the Earth's crust back through deep time.

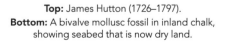

Top: James Hutton (1726–1797).
Bottom: A bivalve mollusc fossil in inland chalk, showing seabed that is now dry land.

KEY **THEORIES** AND **EVIDENCE**

IGNEOUS, METAMORPHIC AND SEDIMENTARY

"WHAT CLEARER EVIDENCE COULD WE HAVE HAD OF THE DIFFERENT FORMATION OF THESE ROCKS, AND OF THE LONG INTERVAL WHICH SEPARATED THEIR FORMATION, HAD WE ACTUALLY SEEN THEM EMERGING FROM THE BOSOM OF THE DEEP."

JOHN PLAYFAIR, 1803

Although the rock cycle is far simpler in its components than the carbon cycle, the cycle itself is significantly more entwined. Rather than forming something like a circle, different types of rock feed into each other through the various processes that cause them to form.

Arbitrarily, we can start with the formation of igneous rocks. These are the result of minerals being pushed down from the upper crust, typically by the movement of tectonic plates, being heated until they melt into the molten magma form, then cooling as they pass back up, often through volcanic systems, where lava emerges on the Earth's surface to cool and form igneous layers. Alternatively, igneous rock can cool "intrusively" (as opposed to extrusively) when it is pushed into other rock types to form an intrusion. These sometimes become visible when the softer rock around them erodes, forming dramatic shapes such as the Devil's Tower in Wyoming, made famous by the movie *Close Encounters of the Third Kind*.

UPPER CRUST

The majority of the Earth's upper crust is igneous. Most common amongst the extrusive igneous rocks is basalt, typically grey to black rock which can be glass-like in fine detail, mostly composed of feldspar and pyroxites (both mixes of metal aluminium silicates). Also common is quartz, a silicon oxide crystalline form. The best-known intrusive form is granite, a hard, granular rock usually featuring a mix of feldspar and quartz.

All types of rock, including igneous, suffer erosion once exposed to the elements, where water and wind gradually remove grains of the rock. These are washed down gravitationally and collect as sediment, joined by debris from organic life, particularly crushed shells from aquatic life. The tiny grains are often linked together initially by a kind of natural glue as minerals are precipitated out of water between the grains, but then compressed as other layers cover them until they form sedimentary rock.

The most familiar forms here are sandstone and limestone (including chalk), though technically coal, for example, can be considered a kind of sedimentary rock. Limestone is typically made up primarily of the remnants of marine life, with shells rich in calcium, resulting in a stone that is mostly calcium carbonate. Sandstones contain silicate crystals, such as feldspar and quartz, but without the melting and reforming of igneous rock, the resultant stone is far softer than basalt or granite.

METAMORPHISM

The final process, metamorphism, takes existing rock and changes its form through extremes of heat and pressure without ever melting it. A large proportion of the Earth's crust is metamorphic (though igneous rock dominates the top layer of the crust). Perhaps the most familiar metamorphic rocks are marble and slate. Marble is limestone or dolomite (calcium magnesium carbonate) that has undergone metamorphosis, while slate starts out as clay or volcanic ash. Although less well known, another common form is gneiss, which can originate both as an igneous rock such as granite or a sedimentary rock such as sandstone.

Each of these types of rock can feed into the others as it works through the rock cycle.

EVIDENCE

The basic concepts of the rock cycle emerged from observation of a wide range of layers of rock and how these appeared to be interrelated, plus studying the outflows of magma from volcanoes and the effects of sedimentation. However, the full modern concept of the rock cycle depended on the understanding of plate tectonics, which was not available until the second half of the twentieth century.

Opposite left: Lava flow on Hawaii producing igneous rock.
Opposite right: The sandstone outcrop of Uluru in Australia.
Above: A marble quarry on Naxos, Greece.

CRITICS

The primary criticism of the rock cycle came from the Neptunists and catastrophists. These were primarily the same individuals, as their idea for the basis of current rock structures was that they were formed during a single catastrophic deluge, often identified with the biblical flood, assumed to have taken place a few thousand years in the past.

Those who supported this approach would point out that, apart from a few volcanoes and earthquakes, the Earth appeared to be stable, running counter to the idea of continued, gradual change. Not all catastrophists equated the deluge they thought responsible for today's rock formations with a worldwide flood – some believed that the cause was less uniform (as the rock record appears to demonstrate), caused by a local catastrophic deluge.

> "NOT ALL CATASTROPHISTS EQUATED THE DELUGE THEY THOUGHT RESPONSIBLE FOR TODAY'S ROCK FORMATIONS WITH A WORLDWIDE FLOOD"

The primary scientific argument against uniformitarianism was that the Earth was simply not old enough to be equipped with the deep time necessary for repeated rock cycles to create the current structures – but from the late nineteenth century into the twentieth century, new techniques supplied evidence for longer and longer lifetimes, providing time in the millions and then billions of years to reach the current estimated age of the Earth at about 4.5 billion years. Some still argue against rock cycles from a religious viewpoint, maintaining the view that most rock structures formed during the supposed year-long period of the great flood around 4,000 years ago, but with no scientific evidence to back up their arguments.

WHY IT **MATTERS**

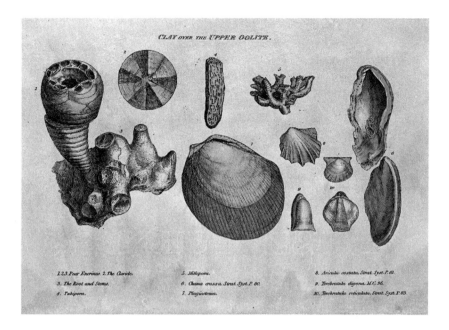

The rock cycle sounds like a dull aspect of geology. Of course, it is important in the sense that it provides the mechanisms by which the various different types of rock came into being – but it is all happening on a timescale far longer than the 200,000 years of human existence, let alone the lifetime of an individual human being. However, there are two very significant aspects of the rock cycle that inform our wider understanding of science.

DEEP TIME

The first is that it gave us confirmatory evidence for the deep time required for the evolution of today's lifeforms to take place. We tend to consider geology and biology separately, but in reality there are many ties. Charles Darwin, for example, bridged the two communities, while the English geologist William Smith compared fossil remains to match up the different layers he found on his progress around the country as a mining engineer. In effect, the rock cycle is not just a geological cycle but a biogeochemical cycle, pulling together the impact of both geological changes and the impact of life on the Earth's crust.

Second, the rock cycle proved crucial to understanding the bigger picture brought in by plate tectonics and our improved understanding of the Earth's structure beneath the crust. It is fundamental to understanding how the different geological structures we live on and interact with were formed.

Opposite left: Great flood fresco in Resurrection Cathedral, Tutaev, Russia.
Opposite right: Neptunists assumed rock formations were produced in the great flood.
Above: Engraving from William Smith's monograph linking strata and fossil types.

FUTURE **DEVELOPMENTS**

Although the rock cycle is reasonably well understood, because it is a complex system – and because the geological timescales are so different from those we experience as lifeforms – there is still more to learn of the detail and to be able to totally describe the mechanisms involved.

Perhaps the most interesting potential developments are not so much in geology as selenology (studying the structure and rocks of the Moon) and aresology (the same role for the planet Mars). For example, feldspar is a common igneous rock form, made up of a range of metal aluminium silicate minerals which collectively form around half of the Earth's crust. From the samples brought back by the *Apollo* missions, we know that feldspar occurs on the Moon (not entirely surprisingly, given current theories on the Moon's origin – see page 227), but also more interestingly that it is present on Mars.

Feldspar has been discovered in Martian meteorites: rocks that were blasted off Mars by an impact from space and eventually found their way to Earth. And in 2012, NASA's Curiosity rover discovered feldspar was a component of Martian sand when for the first time the technique of X-ray diffraction was used on Mars. Our understanding of the rock cycle is beginning to give us a bigger picture that takes in other bodies in the solar system.

Above: Rock samples brought back by the *Apollo* missions allow a study of selenology.

THE **ESSENTIAL** SUMMARY

ORIGINS	KEY THEORIES AND EVIDENCE	CRITICS	WHY IT MATTERS	FUTURE DEVELOPMENTS
1740s Anton Moro proposes the Plutonist theory that rocks were produced by being subject to heat before emerging as magma.	**Igneous rocks** are formed when tectonic plates push existing rocks down to depths where they melt.	Primary criticism came from **Neptunists** and **catastrophists**.	The rock cycle gave **confirmatory evidence** of the **deep time** needed for biological **evolution**.	The rock cycle has **complex interactions** and there is still more to learn.

1770s Abraham Werner puts forward the Neptunist theory of rock structures formed by the influence of water in a catastrophic deluge.

1785 James Hutton publishes his work on rock formations, emphasizing the significance of uniform development.

1830 Charles Lyell publishes *his Principles of Geology*, underpinning Hutton's ideas with observations from volcanoes and an understanding of "deep time".

Igneous rocks can be **extrusive** – emerging as lava from volcanoes – for example, forming **basalt**, or **intrusive** – pushed into other minerals – typically forming **granite**.

All rocks are eroded by water and wind, shedding grains which are washed together and compressed to form **sedimentary rock**.

Sedimentary rocks include **sandstone** and **limestone**, the latter primarily from crushed shells and including **chalk**.

In a **metamorphic rock**, existing rock is transformed under heat and pressure without liquifying. Most familiar are **marble** and **slate**, though **gneiss** is also common.

Evidence for the **rock cycle** came first from observation of rock layers, volcanoes and sedimentation, but the full picture required the theory of **plate tectonics**.

Many catastrophists believed that the current rock formations (apart from local modification by volcanoes) originated in a **catastrophic deluge**, perhaps the **biblical flood**.

It was argued that the Earth was **not old enough** for the rock cycle to provide all the current variety – but with the modern idea of a **4.5-billion-year-old Earth** there was time for many cycles to produce the current structures.

The rock cycle is both **geological** and **biogeochemical**.

The rock cycle was crucial to understanding the **bigger picture** brought in by plate tectonics.

Understanding the rock cycle helps us understand more about the **formation of the Moon and Mars**, as we have begun to explore selenology and aresology as well as geology.

WEATHER SYSTEMS

THE **ESSENTIAL** IDEA

"CLIMATE IS WHAT ON AVERAGE WE MAY EXPECT,
WEATHER IS WHAT WE ACTUALLY GET."

ANDREW HERBERTSON, 1901 (OFTEN ATTRIBUTED TO MARK TWAIN)

The Earth's weather is the outcome of the interaction of a number of systems – the atmosphere, the oceans, the land masses and incoming energy from the Sun. The Sun heats the various components, causing, for example, water to evaporate and rise into the atmosphere where it will form clouds and precipitate as rain, hail and snow. Temperature differences in parts of the atmosphere cause air currents – wind – while the sheer scale of the atmosphere means that the collective interaction of small components can result in lightning, tornadoes and hurricanes.

PREDICTING THE WEATHER

As long as humans have been curious about the world around us, we have looked for ways to predict what will happen with the weather, to make the best of our activities from farming to social events. What started off as a collection of folk methods was transformed into scientific methodology in the nineteenth century. The hope was that predictions would get better and better. But in the second half of the twentieth century it was shown that the weather was a chaotic system in the mathematical sense, and it would never be possible to make long-term predictions.

Our ability to forecast the weather was significantly improved by an approach known as ensemble forecasting, and our knowledge of the mechanisms of the weather continues to improve year on year – but these are systems we will always have to respect in terms of their impact on everyday lives.

ORIGINS

Attempts at weather forecasting came before any understanding of weather systems. Like traditional medicine, early forecasting was a mix of folklore, based on observation, with pure fantasy. So, for example, the weather rhyme "Red sky at night, shepherd's delight – red sky in morning, shepherd's warning" does have some validity because red skies are more common with high pressure in the atmosphere. High pressure is commonly associated with good weather – if observed in the evening it is often moving in, suggesting good weather ahead, while in the morning it implies the high pressure could be on its way out, taking the good weather with it.

MEASURING THE AIR

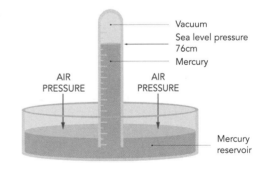

A more direct mechanism for tapping into air pressure, the barometer, was invented in 1643 by Italian scientist Evangelista Torricelli. Early barometers such as Torricelli's were not practical household items, involving as they did a long tube (typically a yard or metre long) full of mercury open to the atmosphere. A key development would be the aneroid barometer, invented by French scientist Lucien Vidi in 1844, which uses movement of a metal cell with a vacuum inside, enabling a simple dial-like instrument to be made.

Scientific forecasts in the modern sense began in the 1850s, produced by Irish scientist Francis Beaufort and English naval officer Robert Fitzroy (also captain of the *Beagle*, which took Darwin on his voyage of discovery) for the British navy. This was followed by the first public forecast in the *London Times* newspaper in 1861. These were predicated on observations from weather stations; mathematical forecasting, based on changes in the state of the weather in three-dimensional chunks of the atmosphere, would start in the 1920s.

The need for better forecasting has driven our understanding of weather systems. Much of this was based on the underlying physics of fluid flow, first given a full scientific description in 1845 when French engineer Claude-Louis Navier's equations were given a physical explanation by English physicist George Stokes, combined with an understanding of radiant heat from the Sun, but it was not until 1961 that the chaotic nature of the weather was first understood by American meteorologist Edward Lorenz.

Vacuum
Sea level pressure 76cm
Mercury

AIR PRESSURE

AIR PRESSURE

Mercury reservoir

Opposite: The Sun's energy is the driver of our weather systems.
Top: Evangelista Torricelli (1608–1647).
Middle: Aneroid barometer.
Bottom: Mercury barometer.

KEY **THEORIES** AND **EVIDENCE**

FORECASTS, MODELS AND CHAOS

The exploration of the nature of weather systems is so closely tied into weather forecasting that there is no real separation possible: it is the need for forecasts that has driven our understanding of the weather.

Forecasts are based on building mathematical models of the world's weather. These models divide the surface of the Earth into cells, which are extended up into the atmosphere, potentially in a number of layers. The models start with observations, plotting the weather conditions in each cell at a starting point in time, then move forward step by step, predicting how each cell will change depending on its neighbours.

A range of factors can come into play in the more sophisticated models. They will include the fluid flow of the atmosphere within each cell, the interchange of heat between cells at different temperatures, the impact of solar radiation, and the changes of state of water entering and leaving the cell. Although early mathematical models were produced by hand, it was only possible to make relatively sophisticated models using a computer. Modern forecasters are amongst the world's biggest users of supercomputers, producing extensive calculations on vast numbers of cells.

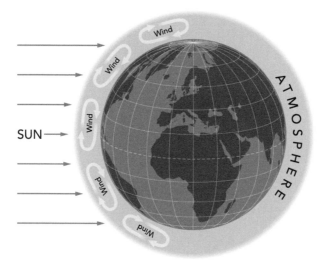

COMPUTER FORECASTING

It was an early attempt to use a computer in forecasting that led to the development of the mathematical theory of chaos and, eventually, a major improvement in the accuracy of forecasts, combined with the understanding that long-range forecasting will never be possible. American meteorologist Edward Lorenz, working at the Massachusetts Institute of Technology, was running a simple mathematical model on an early computer in 1961.

He had previously run the model for part of the period under study. As the computer took hours to produce an output, rather than run it again from the start,

Lorenz input results from a printout as a starting point. He was shocked to discover that the new forecast was soon totally different from the original. It turned out that while the computer itself worked to six decimal places – dealing with numbers, for example, such as 5.724293 – the printout had rounded the numbers to three decimal places, making the equivalent value 5.724.

CHAOS THEORY

Lorenz had discovered that the complex interplay of systems that make up the weather was particularly sensitive to "initial conditions". Very small changes in the way the system started off would result in very different outcomes. Lorenz would typify this with the title of a presentation he made: "Does the flap of a butterfly's wings in Brazil set off a tornado in Texas?" Although his answer to this question was "No" (a tornado is too localized a phenomenon), this so-called "butterfly effect" underlines the particular sensitivity of systems of this kind, which are described mathematically as chaotic.

This sensitivity resulted in a total transformation in the approach to weather forecasting. Rather than trying to make a single, definitive forecast of what would happen, meteorologists started to run their models many times with very small changes in the initial conditions. The outcomes of the runs are combined into an "ensemble forecast" – a typical modern forecast might involve, say, 50 runs of a model. If, for example, rain was predicted for a particular time and place in 20 out of the 50 runs, the model would then predict a 40 per cent chance of rain.

EVIDENCE

Information for forecasting and models has been gathered from a large number of weather stations around the world since the nineteenth century, measuring temperature, air pressure, wind speed, precipitation and more. Originally many of these stations depended on volunteer observers, though now most are automated. Data from the stations has now been joined by the output of a range of technical devices, notably weather satellites, studying weather patterns from above. Mathematical modelling means that it is now possible to see which data is most effective in making predictions, though chaos theory makes it clear that beyond around 10 days, simply saying what the weather tends to be like at that time of year provides a more accurate forecast than any model.

Opposite: The Sun's energy produces temperature differentials, causing wind flow.
Top: Modern ensemble weather forecasts are run many times with slightly different starting conditions.
Middle: The butterfly effect typifies the extreme dependence on small differences.
Bottom: An automated weather station in Austria.

CRITICS

"BUT WHO WANTS TO BE FORETOLD THE WEATHER? IT IS BAD ENOUGH WHEN IT COMES, WITHOUT OUR HAVING THE MISERY OF KNOWING ABOUT IT BEFOREHAND."

JEROME K. JEROME, 1889

It doesn't take long searching online to discover news coverage of "amateur forecasters" who successfully predict the long-term weather outcome better than professionals. What is happening here is a statistical process that is also responsible for apparent expertise in predicting other chaotic outcomes, such as the future of the stock market. The effect is well illustrated by a lottery winner. A lottery winner is apparently a successful forecaster of the outcome – but by focusing on the winner we ignore the millions of people who entered the lottery and didn't win. If a sufficient number of people predict the long-term weather, even by random guesswork, some will get it right. This doesn't mean their method is better than professionals: it's just luck.

The divide between what is possible and what users of forecasts expect is illustrated by US Air Force weather forecasts during the Second World War. An American expert, Kenneth Arrow, analysed forecasts that attempted to predict the weather one month ahead and found they were no better than random guessing – and worse than a guess based on experience. His study was used in an attempt to cease producing these useless forecasts. The Air Force response was, "The Commanding General is well aware that the forecasts are no good. However, he needs them for planning purposes."

Above: Professional forecasts depend on data from many volunteer observers.
Opposite: Hurricane approaching North America.

WHY IT **MATTERS**

Being able to understand the weather sufficiently well to predict how it will progress can have impacts ranging from whether or not a day out is ruined by rain to the life and death outcome of the arrival of a devastating hurricane.

Accurate weather prediction is also of huge importance to those who are growing crops and to those whose travel, particularly by sea or air, who can be significantly impacted by changes in the weather.

> "ALTHOUGH FORECASTERS CAN SOMETIMES PREDICT THESE CHANGES, THE CHAOTIC ASPECT OF WEATHER MEANS THAT THERE IS ALWAYS DOUBT ATTACHED TO A FORECAST"

Hurricanes illustrate well both the importance and difficulty of forecasting. With a minimum speed of 120 kilometres (75 miles) per hour, they are immensely powerful storms that can whip wind speed up to 300 kilometres (190 miles) per hour and wreak havoc when making landfall. Although hurricanes are obvious on weather satellite images – these vast storms can be anything from 30 kilometres (19 miles) to 2,000 kilometres (1,250 miles) across – their paths of destruction are hard to predict as they can suddenly veer off in a new direction, or even double-back on themselves. Although forecasters can sometimes predict these changes, the chaotic aspect of weather means that there is always doubt attached to a forecast, which can result in sudden and unexpected devastation.

Even though weather forecasting is limited by the chaotic nature of weather systems, the process of forecasting has become significantly more accurate since the 1980s, when ensemble forecasts were pioneered.

FUTURE **DEVELOPMENTS**

There is room for improvement in the presentation of weather forecasts. Although an ensemble forecast naturally generates probabilities, these have rarely been adequately explained to the public making use of the forecast. For example, an ensemble forecast might show a value of 40 per cent for rain between 1 and 2 pm. This is often interpreted as there being rain in 40 per cent of locations covered by the forecast, or for 40 per cent of the time. In reality, it means that rain occurred in 40 per cent of the model runs in the ensemble, which gives a prediction of a 40 per cent chance of rain occurring.

As far as understanding the weather goes, we can expect to see better models continue to be developed which take into account more of the subtleties of weather, especially as new satellite technologies are developed. However, this progress is always subject to the proviso that the chaotic nature of the weather systems will never allow perfect prediction.

Above: Weather satellites have greatly improved the ability to track weather patterns.

THE **ESSENTIAL** SUMMARY

ORIGINS	KEY THEORIES AND EVIDENCE	CRITICS	WHY IT MATTERS	FUTURE DEVELOPMENTS
Traditional forecasting was often pure guesswork, but some weather lore (such as "Red sky at night…") does have a basis in observation.	The nature of **weather systems** is inseparable from attempts to **forecast** the weather.	Professional forecasters are often compared unfavourably with **amateurs who make successful long-term predictions**. But this is no more accurate a prediction than is choosing a winning **lottery number**.	Predicting the weather can **avoid a ruined day out** or even **life and death situations** with the arrival of a hurricane.	The public would benefit from **better interpretation of ensemble forecasts** as the use of percentages has often caused confusion.
1643 Evangelista Torricelli invents the mercury barometer.	Forecasts **plot the weather across a series of "cells"** dividing up the surface of the Earth.	In the **Second World War** it was pointed out that **forecasts 1 month ahead were useless**. The response was that they knew they were no good… but **still needed them to be able to plan**.	**Hurricanes** are devastating weather systems, but their **paths are very difficult to predict** as they can suddenly change direction. Even so, **forecasts have got better**.	Models are improving all the time, both mathematically and in the effectiveness of technology from **supercomputers** to **weather satellites**.
1844 Lucien Vidi invents the more practical dial-like aneroid barometer.	Weather models incorporate **atmospheric fluid flow, interchange of heat, solar radiation, water** entering and leaving a cell and more.		Weather forecasting will **always be limited by the chaotic nature of weather systems**, but ensemble forecasts have made a big step forward.	The chaotic nature of weather systems will **never allow perfect prediction**.
1845 George Stokes explains Claude-Louis Navier's equations for fluid flow, central to the understanding of atmospheric flows.	Modern models make use of the **fastest supercomputers** to produce a vast number of sophisticated calculations.			
1850s Scientific forecasts begin for the British navy.	Computer forecasting led to the discovery of **mathematical chaos**. Weather systems are very sensitive to **tiny variations in initial conditions**, making them impossible to forecast long term.			
1861 The first public forecast is published in the *London Times* newspaper.	Modern weather forecasts are based on an **"ensemble"**: the model is run many times with small changes in initial conditions and the **probabilities of different outcomes** calculated.			
1920s Mathematical forecasting is first used.				
1961 Edward Lorenz discovers the chaotic nature of weather systems.				

CLIMATE CHANGE

THE **ESSENTIAL** IDEA

"THE DECISIONS WE MAKE TODAY ARE CRITICAL IN ENSURING A SAFE AND
SUSTAINABLE WORLD FOR EVERYONE, BOTH NOW AND IN THE FUTURE."

DEBRA ROBERTS (IPCC WORKING GROUP), 2018

Climate is often confused with weather – weather is what we experience locally;
climate is the global whole producing those effects. Climate systems include the
atmosphere, the oceans, the planet's land surface and items which interact with
these, such as clouds, plants and animals, notably human beings.

The Earth's climate has changed throughout history, ranging from deep ice ages
to the dominance of tropical conditions. Technically, we are currently in an ice age,
though in an "interglacial" period when the ice sheets have withdrawn. The most
significant and controversial factor influencing the climate is the greenhouse effect.

This effect is the result of certain gases in the atmosphere, which let in light
from the Sun, warming the Earth, but prevent heat from escaping. Since the

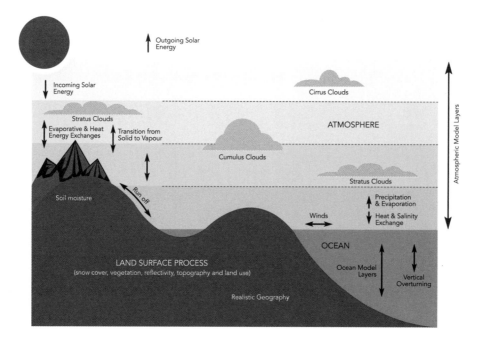

industrial revolution, the level of these gases in the atmosphere has increased, and there is strong evidence that man-made greenhouse gases increase global temperature. The impact on the climate is complex: winters are colder as well as summers hotter. Climate change also increases the power of storms and increases sea level. There is scientific consensus that climate change is real and having a negative impact on many lives.

ORIGINS

Although based on limited evidence, as far back as Ancient Greece it was thought that the climate of a region changed over time. As geology developed, giving a view into the past through the way that rocks had been laid down and distorted, it became clear that there had been major climate changes in the past – for example, based on the action of glaciers in now-temperate regions, shaping valleys and depositing large rocks that would not otherwise be found in their present location.

By the 1830s, although a contingent still believed that many unusual features had been caused by the biblical flood, the idea of "ice ages" developed by Swiss geologist Louis Agassiz was gaining traction. The causes of such huge variations in climate were unknown, though over time it became clearer that a combination of greenhouse gases in the atmosphere, solar cycles where the Sun's output changed, and shifts in land masses through continental drift all contributed in the past.

EARLY CLIMATE SCIENCE

The idea of climate change driven by greenhouse gases feels like a modern concept, but was put forward by Swedish physicist Svante Arrhenius in 1896. Surprisingly, it was observations of the Moon that enabled Arrhenius to measure levels of the greenhouse gas carbon dioxide (CO_2) in the atmosphere. Carbon dioxide absorbs infra-red light. By comparing data on the infra-red content of moonlight depending on whether the Moon was high or low in the sky (when lower, its light passes through more of the atmosphere), Arrhenius was able to calculate how carbon dioxide influenced warming and cooling and hence what an effect changes in CO_2 levels had on the planet. The major concern at the time was that greenhouse gas levels would drop. Arrhenius believed that it would only take a reduction of about 100 parts per million of atmospheric CO_2 – halving levels at the time – to trigger the return of a full-scale ice age.

Researchers at the time were also aware that carbon dioxide levels were gradually increasing due to the impact of steam engines and other uses of fossil fuels, but (being pretty much universally inhabitants of chilly northern climes) they regarded a mild increase in global temperature as a boon. It was not until 1972, with the publication of a study by English meteorologist John Sawyer, that effective modelling of global warming and its impact began.

Opposite: The climate reflects the balance of a complex interplay of factors.

KEY **THEORIES** AND **EVIDENCE**

GREENHOUSE EFFECT, MODELS AND IMPACT

Central to climate change is the greenhouse effect. This is produced by gases in the Earth's atmosphere interfering with the flow of energy away from the Earth. The primary source of our energy is the Sun. Although the Earth has some internal heating, the majority of the energy that keeps the Earth inhabitable comes from the Sun as electromagnetic radiation, notably visible light.

This light warms up the surface of the Earth. Anything warm also gives off electromagnetic radiation – at surface temperatures, this is primarily infra-red. As the infra-red radiation streams away from the Earth's surface, some gas molecules in the atmosphere absorb the energy, then re-emit it. As a result, some of the radiation is now heading back to Earth – the greenhouse gases act to keep heat in.

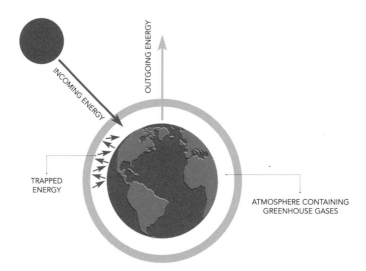

We need to be grateful for this effect. Without it, the average temperature on the Earth would be 0°F (-18°C) – around 60°F (33°C) below current levels. Liquid water would be uncommon and life as we know it would almost certainly never have developed. However, it's also true that too much greenhouse gas in the atmosphere is a problem, as it only takes a few degrees of warming to have a major impact on sea levels, weather extremes and wildfires.

There are a number of greenhouse gases – water vapour, methane and nitrous oxide, for example – but the best-known is carbon dioxide, produced by burning carbon-based fuels, such as coal, wood, oil and petrol. By the 1970s, some scientists were already predicting that increases in levels of greenhouse gases would have a significant impact on the climate. Although there were initially some concerns of global cooling caused by pollution in the atmosphere, by the 1980s there was a broad scientific consensus forming that the biggest impact on the climate would come from growing levels of greenhouse gases – levels that were not just increasing, but accelerating as more countries adopted fully industrial economies. These levels are now twice what they were at the start of the industrial revolution.

MODELLING THE CLIMATE

There is no scientific doubt that rising greenhouse gas levels are having a noticeable impact on the climate. Where there is disagreement is the rate at which these effects will progress in the future. Weather systems are complex, chaotic and impossible to forecast successfully more than a few days ahead. Climate forecasting is easier, because it does not attempt to predict what will happen in a small area, but still requires many factors to be projected decades into the future. Such models also have to take into account amplifying and feedback effects. For example, as ice melts it makes the Earth's surface darker, which means it warms up more from incoming light, causing more ice to melt. However, there is agreement that unless global emissions are capped, the impact of climate change will get much worse.

Amongst significant impacts are a rise in sea level, hotter summers and colder winters, with resultant flooding, poor harvests, heat deaths, wildfires and more. A rise in sea level is a double whammy: as the Earth's temperature increases, seawater expands, so the levels rise – but also ice sheets melt into the oceans, further pushing up the levels. Left unchecked, climate change could result in sea levels rising sufficiently to wipe out many coastal cities.

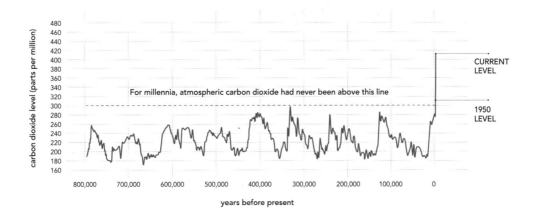

EVIDENCE

Levels of greenhouse gases have been monitored for a number of decades, and we are able to measure levels further back using ice cores. These are samples drilled out of glaciers and ice sheets. Over the years, accumulating layers of ice capture bubbles of air, making it possible to look back in time to see how levels have changed. CO_2, for example, is now at levels that have not been seen for thousands of years.

At the same time, global temperatures continue to increase. We are now beginning to see impacts from this. Sea levels are rising at around 30 centimetres (1 foot) per century. This doesn't sound much, but the rate is accelerating – and even a few inches of rise means that tidal extremes, causing coastal flooding, increase significantly. Although scientists are reluctant to say that a specific extreme weather event is caused by climate change, we are now seeing these occur with increasing frequency. It is not possible to doubt that climate change is having a damaging effect on many lives.

Opposite: Greenhouse gases trap energy that would otherwise radiate into space.
Above: Carbon dioxide levels have been transformed since the industrial revolution.

CRITICS

"THE WHOLE CLIMATE CRISIS IS NOT ONLY FAKE NEWS,
IT'S FAKE SCIENCE."

INDUSTRIAL CONSULTANT PATRICK MOORE, QUOTED IN A TWEET BY US PRESIDENT DONALD TRUMP

There are vocal critics of climate change science. Much of the criticism has been driven by industries facing reduction in revenues from lower emissions, though a few climate sceptics are established scientists who doubt climate models. One strong theme is that the problem is not being treated appropriately economically, notably from Danish author Bjørn Lomborg, who argues that efforts to prevent climate change are influenced by emotion rather than logic, arguing we would be better focusing resources on issues such as malaria and safe water supplies. Lomborg has commented, "The solution to climate change cannot be found in personal changes in the homes of the middle classes of rich countries." Instead, he suggests, we should implement a carbon tax and focus on innovation, spending more on green energy.

COLD STARS

An example of a scientific alternative to human-caused climate change is the "chilling stars" theory of physicist Henrik Svensmark. This suggests that cosmic rays, streams of high-energy particles which hit the atmosphere from deep space, trigger cloud formation, and that changes in the Sun's magnetic field influence how many cosmic rays get through. As a result, he suggests, they influence the climate because low clouds cool the Earth. It's true that clouds have an impact on temperature (though, confusingly, high clouds warm the Earth), but there is no good evidence for this effect having a significant impact. There is now a better scientific agreement on climate change than most other scientific theories – and the majority of scientists opposing the consensus are not climate experts.

Above: Bjørn Lomborg (b1965).
Opposite: Wildfires have become increasingly common with global warming.

WHY IT **MATTERS**

Climate change will have a huge impact on human lives. Take two examples: sea levels and wildfires. Low-lying regions are easily devastated by sea level rise. In a storm surge in 1998, around 65 per cent of Bangladesh ended up under water. Many major cities – New York and London, for example – are at risk from sea level rise. Equally, in 2019, wildfires in Australia, California, Indonesia, Africa and the Amazon had impacts ranging from hundreds of thousands of evacuations to increased deaths from air pollution, destruction of homes and direct loss of life.

The longer-term impact of climate change could be to render whole areas uninhabitable, creating a refugee crisis. There is no doubt that worldwide protests and media coverage raise awareness of climate change, though sometimes this publicity can be counter-productive, where over-dramatization turns listeners against the message. However, we need to take faster action to reduce greenhouse gas emissions and to remove gases from the atmosphere.

A significant problem is that climate change is not localized. Individual actions – such as flying less, switching to electric vehicles and public transport, changing home heating and eating less meat – though positive, have a trivial impact. It is only if governments, particularly of the biggest emitters (the top four are China, the USA, India and Russia), can agree on significant changes that serious damage can be avoided.

FUTURE **DEVELOPMENTS**

The science of climate change is well established. We need to reduce our greenhouse gas emissions quickly and to develop technologies to remove greenhouse gases from the air.

In part this will involve a move away from fossil fuels (coal, oil, gas and petrol) to electricity generated from non-emitting sources: solar, wind, hydroelectric, tidal power and nuclear. This is happening, but costs are loaded against non-emitting energy, with, for example, the costs of electric cars still far higher than fossil fuel vehicles. Battery technology is developing fast, and the differential will disappear – but there are concerns that we don't have sufficient time available.

"AS YET, RELATIVELY LITTLE WORK HAS BEEN DONE ON TECHNOLOGIES TO REMOVE GREENHOUSE GASES FROM THE ATMOSPHERE"

As yet, relatively little work has been done on technologies to remove greenhouse gases from the atmosphere, but these will be essential to meet realistic targets. An example of the technology involved is a new material a thousand times more efficient at removing CO_2 from the atmosphere than trees. The resin, developed at Arizona State University's Center for Negative Carbon Emissions, absorbs carbon dioxide from the air when dry and releases it when wet. Panels of the material could be used to soak up CO_2, which would then be taken to locations where the gas can be stored underground.

Above: The move away from fossil fuels is only a part of the response required by climate change.

THE **ESSENTIAL** SUMMARY

ORIGINS	KEY THEORIES AND EVIDENCE	CRITICS	WHY IT MATTERS	FUTURE DEVELOPMENTS
1837 Louis Agassiz publishes the concept of ice ages. **1896** Svante Arrhenius shows the impact of greenhouse gases on global temperatures. **1972** John Sawyer makes first detailed prediction of global warming.	**Greenhouse effect:** some atmospheric gases trap infra-red radiation leaving the Earth and re-emit it back, warming the surface. Without greenhouse gases, Earth would be too cold for life as we know it, but increasing levels since the industrial revolution threaten the global climate. **Main greenhouse gases**: water vapour, carbon dioxide, methane and nitrous oxide. **Impacts**: sea level rise, hotter summers and colder winters, with resultant effects including flooding, poor harvests, heat deaths, wildfires and more. **Evidence:** increases in gases can be tracked back many centuries from air trapped in ice cores. Levels have been rising since the industrial revolution and are now higher than they have been for many thousands of years. Global temperatures are consistently rising, and the rate of rise is accelerating.	**Some deny climate change** due to vested interests, such as industries facing reduction in revenues and increased costs. **Bjørn Lomborg**, since publishing his book *The Skeptical Environmentalist*, has stressed a logical economic view to climate change, arguing for carbon tax and a focus on innovation, notably in green energy generation and storage. **Henrik Svensmark,** a physicist, suggests that climate change could be caused by the impact of changes in the solar magnetic field on cosmic rays, which might change cloud cover. A consensus of climate scientists agrees that climate change is real and human-produced greenhouse gases are primarily responsible.	It is already having **significant impact** on human lives. Outcomes such as **sea level rise and wildfires** have devastating impacts on communities. Longer-term impacts could **render whole areas uninhabitable**, triggering refugee crises. Action at a personal level has limited impact. Only **governmental changes** can make a major difference.	Energy production needs to move away from fossil fuels to sources that do not generate greenhouse gases, such as **solar, wind, wave, hydroelectric and nuclear**. Improved battery technologies will make **electric vehicles** more affordable – but it is taking too long. More work needs to be done on technologies to **remove carbon dioxide from the atmosphere**. An example is a resin that is 1,000 times better at removing carbon dioxide from the atmosphere than trees.

KEY PEOPLE, IDEAS AND CONCEPTS

KEY SCIENTISTS

ANCIENT

c500–450 BCE **Leucippus** co-originator of atomic theory

c490–430 BCE **Empedocles** originator of the theory of elements

c460–370 BCE **Democritus** co-originator of atomic theory

384–322 BCE **Aristotle** enhanced theory of elements, early concepts of physics and biology

287–212 BCE **Archimedes** mathematician and engineer

MEDIEVAL

c965–1038 **Hasan Ibn al-Haytham** developed basic concepts of optics

c1170–1250 **Leonardo of Pisa (Fibonacci)** described the Fibonacci sequence and popularized Arabic/Indian numerals in Europe

1473–1573 **Nicolaus Copernicus** put the Sun at the centre of the universe

1494–1555 **Georgius Agricola** made first scientific study of mining and ores

1544–1603 **William Gilbert** investigated magnetism

1564–1642 **Galileo Galilei** explored the physics of motion and supported Copernican theory

1596–1650 **René Descartes** developed analytical geometry and dualism

ENLIGHTENMENT AND VICTORIAN

1627–1691 **Robert Boyle** made the transition from alchemy to chemistry

1629–1695 **Christiaan Huygens** made astronomical discoveries, worked on gravitation, invented the pendulum clock

1632–1723 **Antonie van Leeuwenhoek** discovered bacteria

1635–1703 **Robert Hooke** worked on the physics of motion, optics and named the biological cell

1643–1727 **Isaac Newton** developed laws of motion, theories of gravity, theory of colour

1707–1778 **Carl Linnaeus** introduced the modern
binomial naming system for organisms

1727–1797 **James Hutton** transformed geology
with the uniformitarian approach

1743–1794 **Antoine Lavoisier** introduced the basics of modern chemistry

1766–1844 **John Dalton** developed modern atomic theory

1769–1832 **Georges Cuvier** introduced basic
biological classification and paleontology

1791–1867 **Michael Faraday** developed electromagnetic and field theories

1796–1832 **Sadi Carnot** explained the physics of heat engines

1797–1875 **Charles Lyell** introduced modern geological theory

1809–1882 **Charles Darwin** developed the theory
of evolution by natural selection

1822–1884 **Gregor Mendel** introduced the basic concepts of genetics

1824–1907 **William Thomson** (Lord Kelvin)
developed the theory of thermodynamics

1831–1879 **James Clerk Maxwell** introduced statistical
mechanics, developed equations of electromagnetism

1834–1907 **Dimitry Mendelev** developed the periodic table of the elements

1844–1906 **Ludwig Boltzmann** worked on kinetic
theory of gases and on entropy

MODERN

1856–1940 **J. J. Thomson** discovered the electron

1858–1947 **Max Planck** introduced quantum physics

1867–1934 **Marie Curie** developed science of radioactivity

1871–1937 **Ernest Rutherford** discovered the atomic
nucleus and co-discovered isotopes

1879–1955 **Albert Einstein** developed special and general theories of
relativity, helped found quantum theory, proved the existence of the atom

1880–1930 **Alfred Wegener** developed the theory of continental drift

1882–1935 **Emmy Noether** devised the mathematics
linking physics and symmetry

1885–1962 **Niels Bohr** developed the quantum theory of the atom

1887–1961 **Erwin Schrödinger** developed the
central equation of quantum physics

1901–1976 **Werner Heisenberg** developed the uncertainty principle

1916–2004 **Francis Crick** co-discoverer of the structure of DNA

1917–2008 **Edward Lorenz** developed chaos theory,
explaining the behaviour of weather systems

1918–1988 **Richard Feynman** developed quantum electrodynamics

1928– **James Watson** co-discoverer of the structure of DNA

1928–2016 **Vera Rubin** put forward evidence for the existence of dark matter

FURTHER READING

PHYSICS AND COSMOLOGY

Our Mathematical Universe – Max Tegmark (Penguin, 2015)
Gives the basics on modern cosmology, but extends from here to some of the more speculative aspects of the field, bringing in the relationship between mathematics and reality.

QED – Richard Feynman (Penguin, 1990)
Taken from lectures for the public on quantum electrodynamics, not the easiest read, but a masterclass in this central theory of quantum physics.

The Beginning and End of Everything – Paul Parsons (Michael O'Mara, 2018)
An impressively in-depth trip through cosmology from the Big Bang to the end of the universe, taking in the formation of stars and galaxies, black holes, dark matter, dark energy and more.

The Quantum Age – Brian Clegg (Icon Books, 2015)
Combines a dive into more depth on the nature of quantum physics with the remarkable stories of the development of the quantum technology that is so important to the modern age.

The Reality Frame – Brian Clegg (Icon Books, 2017)
Provides a ground-up explanation of Galilean, special and general relativity, tying it in to our place in the universe.

The World According to Physics – Jim Al-Khalili (Princeton University Press, 2020)
In a compact book, Al-Khalili provides a surprisingly in-depth exploration of the "three pillars" of modern physics: relativity, quantum theory and thermodynamics.

CHEMISTRY

Periodic Tales – Hugh Aldersey-Williams (Viking, 2011)
Rather than take a structured tour of the elements, the book makes use of artistic and cultural associations to link together stories of the science behind these icons of chemistry.

The Disappearing Spoon – Sam Kean (Little, Brown, 2010)
Engaging tales of the chemical elements, diving into their discovery, use and general oddity. An entertaining combination of fascinating facts and strange people.

The Periodic Table – Eric Scerri (OUP, 2019)
Scerri is the world's leading expert on the development of the periodic table, and though this book can get into a lot of detail, he manages to keep the topic readable and engaging.

BIOLOGY

Dry Store Room No. 1 – Richard Storey (Knopf Publishing, 2008)
Uses the stores of London's Natural History Museum to explore natural history and paleontology.

I Mammal – Liam Drew (Bloomsbury, 2017)
A straightforward biology book on mammals and their origins – but written as a delightful voyage of discovery, piling in surprising facts.

Making Eden – David Beerling (OUP, 2019)
A rare title on the Cinderella of biology, botany. Lots of detail – there's a whole chapter on stomata, plants' gas valves – yet manages to make the topic remarkable reading.

The Accidental Species – Henry Gee (Chicago University Press, 2014)
Provides an introduction to paleontology and a fresh and effective look at the nature of evolution and the development of the human species.

What Do You Think You Are? – Brian Clegg (Icon Books, 2020)
Explores what makes you the individual you are, from the atoms that make you up to your genes, personality and environment.

The Story of the Dinosaurs in 25 Discoveries – Donald Prothero (Columbia University Press, 2019)
Combines stories of the history of paleontology with information on a wide range of dinosaurs in a highly accessible fashion.

EARTH

Atmosphere of Hope – Tim Flannery (Penguin, 2015)
After taking us through the evidence that climate change exists and the predictions of dire future outcomes, Flannery provides a good picture of the possible solutions and their implications.

Jet Stream – Tim Woolings (OUP, 2019)
Starting on a beach in Barbados, takes us on a trip alongside a weather balloon, around the world on this dramatic flow in the global atmosphere with a significant impact on climate.

The Planet in a Pebble – Jan Zalasiewicz (OUP, 2010)
Starting from a single slate pebble, found on a Welsh beach, the author reveals a whole host of aspects of the formation of the universe, the Earth, early biology and geology.

The Story of Earth – Robert M. Hazen (Viking, 2012)
A highly interesting read and a great introduction to modern geological thought on the formation of our planet, both where we came from and where we might be going.

The Vital Question – Nick Lane (Profile Books, 2015)
A fascinating guide through the magnificent complexity of biological mechanisms, particularly those of molecular machinery, leading to an exploration of the possible origins of life.

GLOSSARY

Absolute zero – the lowest possible temperature, –273.15°C, about –460°F.

Accretion – material gathering together under the influence of gravity, responsible for the formation of the solar system.

Angular momentum – a measure of the strength of rotation (and difficulty of stopping it).

Anthropic principle – logical argument based on the existence of human life. For example, some constants of nature must have values in a limited range, or we would not be here to observe them.

Antimatter – all matter particles have equivalent antimatter particles with opposite values for electrical charge and some other properties.

Apex predator – the top of the food chain in a particular environment, which does not have significant predators of its own.

Archaea – though resembling bacteria, these single-celled prokaryote organisms are a totally separate domain with unique properties.

Asteroid – a rocky body in a solar system, smaller than a planet and not a moon.

Atom – basic particle of an element. Once thought to be the smallest possible unit, but now known to have an internal structure.

Bacteria – domain of single-celled prokaryote organisms – the most common form of organism on the Earth.

Black body – a hypothetical object that absorbs all electromagnetic radiation. A heated black body gives off radiation with a specific distribution of frequencies.

Black hole – a collection of matter that has become so dense that it collapses to a point, usually as a result of a stellar explosion. Near the black hole, spacetime is so warped that even light cannot escape.

Blueshift – when a light-emitting object is moving towards the observer, the light becomes higher energy and shorter wavelength, shifting towards the blue end of the spectrum.

Bond (chemical) – electromagnetic link between atoms. Can be covalent, based on shared electrons, ionic based on atoms becoming ions and hence electrically charged, or rely on relative charge because of the shape of a molecule.

Bosons – fundamental particles that act as force carriers, such as photons of light, many of which can occupy the same position and state simultaneously.

Carrying capacity – the population of a species that an environment can support.

Catalyst – a chemical element or compound that makes other chemicals react faster, or at a lower temperature, without itself being consumed.

Chaos – in the mathematical sense, a system where small changes in starting conditions result in major differences in outcome.

Chlorophyll – green pigments found in plants, algae and some bacteria, which are used to absorb energy from light.

Chloroplast – structures within the cells of plants that process energy from light that has been absorbed by chlorophyll so that the plant can make use of it.

Chromosome – single strand of DNA within a cell that contains genes. A cell may have a considerable number of different chromosomes.

Cladistics – a biological system of classifying organisms into groups that have a common ancestor.

Conservation law – physical law where some aspect of a system cannot change if the system is isolated, such as conservation of energy.

Cosmic rays – streams of high-energy particles that flow through the universe, some of which pass into the Earth's atmosphere.

Cosmology – the study of the cosmos, the universe as a whole, as opposed to astronomy, which covers the individual components such as galaxies, stars and planets.

CRISPR (Clustered regularly interspaced short palindromic repeats) – specific DNA sequences that can be used in a precision mechanism to edit genes.

Dark energy – the unknown energy that is accelerating the expansion of the universe, which amounts to around 68% of the mass/energy in the universe.

Dark matter – unknown type of matter that does not interact electromagnetically which influences the rotation of galaxies. Amounts to around 27% of the mass/energy of the universe.

Decoherence – the tendency of quantum particles to interact with their surroundings and lose their state of quantum superposition.

DNA (Deoxyribonucleic acid) – molecules used in living cells to store the genetic information required for organisms to grow and reproduce.

Electromagnetism – force of nature combining electricity and magnetism, responsible for most everyday interactions of matter and for light.

Electron – fundamental matter particle with a negative electrical charge, carrying electrical current. Their distribution around an atom specifies its chemical properties.

Element – material made from a single type of atom. The specific element is determined by the number of particles in its nucleus, while its chemical properties are determined by its outer shell of electrons.

Emergent – a system has emergent properties if they are not properties of parts of the system but rather emerge from the interaction of its components.

Endosymbiosis – the absorption of one or more organisms into another, providing services to the outer organism in return for safety and nutrition, resulting over time in the formation of a compound organism.

Energy – the capability to make things happen, depending on motion or potential from forces of nature such as gravity and electromagnetism.

Entanglement – a linkage between two or more quantum particles that means that a change in one particle is instantly reflected in the other(s), however far they are separated.

Entropy – measure of the disorder in a system. The second law of thermodynamics says that in a closed system, entropy will stay the same or increase.

Epigenetics – influences on genetic mechanisms that come from outside the genes, either from other parts of the DNA or from external additions to the DNA.

Equilibrium – the balance of a system. If a system is in equilibrium it stays in the same state, where out of equilibrium, often as a result of a transfer of energy from one place to another, it changes.

Eukaryotes – organisms based on one or more complex cells which contain a nucleus holding DNA and molecular machinery.

Fermions – fundamental matter particles, such as quarks and electrons. No two fermions can be in the same location with identical quantum properties.

Flagellum – external whip-like structure used like a propeller to move some single-celled organisms.

Field – a natural quantity that has a value across space (and often time), such as the electromagnetic field or a gravitational field.

Force – something acting on an object that will cause its motion to change unless it is resisted.

Fossil fuel – fuel based on carbon from fossilized organisms, such as coal, oil and gas.

Frequency – the number of times a repeating phenomenon, such as a wave, repeats in one second.

Gene – a section of DNA that contains the information to specify an amino acid, usually as a component for a protein.

Gluon – a boson that is the fundamental particle carrying the strong nuclear force that binds together the particles in an atomic nucleus.

Greenhouse gas – gas in the atmosphere that lets heat in, but restricts it from leaving, such as carbon dioxide or water vapour.

Half-life – the time in which half of a quantity of radioactive material will decay.

Hydrogen bond – bond formed by the attraction between the relatively positive end of a molecule and the relatively negative end of another, typified by the hydrogen–oxygen attraction in water.

Inflation – hypothetical sudden expansion of the very early universe, far faster than the speed of light.

Interglacial – period in an ice age when the ice has partially withdrawn. We are currently in an interglacial.

Ion – an atom that has lost or gained one or more electrons to become electrically charged.

Isotope – variant of an element with different numbers of neutrons in the nucleus: isotopes are chemically identical but differ in tendency to decay.

Mantle – section of the Earth lying between the outer crust and the inner core.

Mass – property of a piece of matter that resists acceleration and determines the strength of its gravitational attraction.

Mechanics – the science of movement of objects.

Median – unlike the usual average (the mean) where all values are added together and divided by the number of values, the median is the midway point of the values.

Mitochondria – interior components of eukaryotic cells that store energy and were once thought to be independent bacteria.

Molecule – atoms linked together by chemical bonds to form a single entity.

Multiverse – hypothetical cosmos consisting of more than one universe, each of which could have its own Big Bang and own natural laws.

Neutrino – fundamental matter particle produced in nuclear reactions with very low mass and no electromagnetic interaction.

Neutron – electrically neutral compound matter particle found in the nucleus of atoms.

Nuclear fission – the splitting of an atomic nucleus, generating energy: power source of current nuclear power stations.

Nuclear fusion – the joining of two or more atomic nuclei, generating energy: power source of stars and future nuclear power stations.

Nucleus (biology) – section within a eukaryotic cell separated off by a membrane, containing DNA and molecular machinery.

Nucleus (physics) – relatively massive central part of an atom, made up of protons and neutrons, themselves made up of quarks and gluons.

Orbit – the path a celestial body takes when moving around another body under gravitational attraction.

Orbital – the probability pattern around an atomic nucleus indicating where an electron is likely to be found.

Oxidation – in chemistry, a reaction in which an atom or molecule loses electrons.

Panspermia – theory that life did not originate on Earth but arrived from space.

Parallax – apparent motion of a distant object when observed from two separate locations.

Particle – small quantum component of matter or energy.

Phage – type of virus that attacks bacteria: short for bacteriophage.

Photon – quantum particle of light.

Photosynthesis – the process by which plants, algae and some bacteria generate chemical energy from light.

Plasma – a gas-like collection of ions.

Power – the rate at which work is done (energy is transferred).

Prokaryotes – single-celled organisms with relatively simple cell structure and no nucleus: bacteria and archaea.

Protein – organic chemical compound in the form of a large molecule built from chains of amino acids, central to the functioning of life.

Proton – electrically positively charged compound matter particle found in the nucleus of atoms.

Quantum – having a nature that is divided into chunks rather than continuous: properties of quantum particles can only have certain values, rather than any value.

Quark – fundamental matter particle, three of which make up protons and neutrons.

Radiation – either electromagnetic radiation (such as light) or streams of high-energy particles such as alpha or beta particles, produced in nuclear reactions.

Redshift – when a light-emitting object is moving away from the observer, the light becomes lower energy and longer wavelength, shifting towards the red end of the spectrum.

Reduction – in chemistry, a reaction in which an atom or molecule gains electrons.

Refraction – the result of light passing from one material into another where it has a different speed, resulting in a change of direction.

Relativity – the study of the relative motion between different bodies and the effects this has on time, space and gravitation.

Rewilding – the concept of returning animals and plants to an environment, or replacing extinct organisms with an equivalent, in an attempt to rebalance the ecology.

RNA (Ribonucleic acid) – a simpler equivalent of DNA used to carry genetic information in some organisms, and from place to place when genes are being read and manipulated.

Shell (atomic) – the region around an atom occupied by one or more electrons with some properties in common.

Solar wind – spray of energetic particles from the Sun.

Spacetime – unifies the concept of space and time, which cannot be considered separate under the special and general theories of relativity.

Spectroscopy – making use of the colours of light given off or absorbed by a material or object to determine the elements that make it up.

Spectrum – the range of colours of light emitted or absorbed by a body.

Subduction – process taking place where tectonic plates meet and one plate is forced under the other.

Superconductivity – the ability of some materials at low temperatures to conduct electricity with no resistance.

Supernova – stellar explosion which concentrates energy sufficiently to generate heavier elements and can form ultra-dense bodies such as black holes.

Superposition – a quantum particle or system's ability to be in more than one state at the same time.

Tectonic plate – section of the Earth's crust that very gradually moves over the surface.

Thermodynamics – the science of the movement of heat that takes in the nature of entropy and the conservation of energy.

Uniformitarianism – geological theory that many geological structures are produced over a long timescale by gradual processes that are still under way.

Valence – specifies the ways that an atom will bond with other atoms, dependent on the number of electrons in its outer shell.

Virus – organism that can only replicate making use of the mechanisms of another organism, usually damaging its host in the process.

Wavelength – the distance in a repeating process such as a wave between two identical points in the process.

Work – energy being put to use to make something happen.

INDEX

CREDITS

The publishers would like to thank the following sources for their kind permission to reproduce the pictures in this book.

Key: t: top, b: bottom, l: left, r: right & c: centre

Alamy: 11t, 11b, 19b, 22, 23, 27t, 43t, 43b, 45, 51t, 59, 67b, 71, 75t, 78, 91t, 94r, 120r, 125b, 125c, 133t, 133b, 146, 151, 157b, 159b, 160r, 171t, 187, 193, 195, 198, 208, 213, 215t, 217t, 217c, 219, 220, 224, 225, 228l, 228r, 233t, 233b, 242, 245, 249t, 252, 254, 257b, 259, 260l, 267b, 268; /The American Institute of Physics: 170; /D. Roddy_Barringer: 95b; /Courtesy of Science History Institute: 149t; /Getty Images: 7, 75b, 94l, 110b, 122, 141l, 160l, 191r, 194, 207t, 249b, 265c; /Bruno Gilli/ ESO: 91b; /Huntington Digital Library: 83b; /Photo by Emil Jupin: 276; /Library of Congress: 67t, 83t, 133c; /LSST Project Office: 104; / NASA: 79, 85, 96, 100, 102, 230, 246, 262; /Private Collection: 101t, 175t, 207b, 225t, 267, 273; /Public Domain: 117t, 136r, 141br, 157t, 159t, 175b, 184, 191c, 241, 257t, 260, 265t; /Science Photo Library: 14, 15, 86t, 95b, 99t, 99b, 112t, 129; /Shutterstock: 10, 12, 13, 16, 19t, 27b, 30, 32, 34l, 34r, 35t, 35b, 39t, 39b, 46, 48, 51c, 54, 58, 63, 87, 91b, 103, 106, 108, 110t, 111, 112b, 121, 124, 125t, 127, 137, 141r, 148, 149b, 152, 153l, 153r, 154, 156, 158l, 158r, 161, 162, 166, 167t, 167b, 168t, 168b, 169, 171b, 172, 174, 176, 177, 178, 179, 182, 183t, 183b, 185b, 188, 190, 191l, 200, 202t, 202b, 203, 204, 206, 211, 214, 217c, 218t, 218b, 226t, 229, 233c, 235, 236, 237t, 237b, 238, 244, 250, 251, 253t, 253b, 258l, 259r, 260r, 264, 267tc, 269, 270, 277; /Smithsonian Open Access: 120l; /University of Queensland: 128; /Mark Ward: 101b; / Weizmann Institute of Science: 102t; /Wellcome Images: 117c, 136l, 199t, 215b; /Wikimedia Commons: 226; /Amble 86b; /Chiswick chap 185t; / Gretarsson 243; /Gregory H. Revera 227; /Jeroen Rouwkema 199; /Wang Xiaoli: 180

Every effort has been made to acknowledge correctly and contact the source and/or copyright holder of each picture and Welbeck Publishing apologises for any unintentional errors or omissions, which will be corrected in future editions of this book.